Remote Sensing and Urban

Remote Sensing and Urban Analysis

Edited by

Jean-Paul Donnay
Department of Geomatics
University of Liège

Michael J. Barnsley
Department of Geography
University of Wales, Swansea

Paul A. Longley
Department of Geography
University College, London

GISDATA 9

SERIES EDITORS
IAN MASSER and FRANÇOIS SALGÉ

CRC Press
Taylor & Francis Group
Boca Raton London New York

CRC Press is an imprint of the
Taylor & Francis Group, an **informa** business

CRC Press
Taylor & Francis Group
6000 Broken Sound Parkway NW, Suite 300
Boca Raton, FL 33487-2742

First issued in paperback 2019

ISBN-13: 978-0-7484-0860-3 (hbk)
ISBN-13: 978-0-367-39774-6 (pbk)

British Library Cataloguing in Publication Data
A catalogue record for this book is available from the British Library

Library of Congress Cataloging in Publication Data
A catalog record for this book has been requested

Visit the Taylor & Francis Web site at
http://www.taylorandfrancis.com

and the CRC Press Web site at
http://www.crcpress.com

Contents

III FROM URBAN LAND COVER TO LAND USE 69

List of Figures

List of Tables

List of Contributors

Hans-Peter Bähr   Institut Photogrammetrie und Fernerkundung,
                         Universit Karlsruhe (TH),
                         Engelerstrasse 7,
                         Postfach 69 80,
                         7500 Karlsruhe 1,
                         Germany.

Michael J. Barnsley   Environmental Modelling and Earth Observation Group,
                         Department of Geography,
                         University of Wales Swansea,
                         Singleton Park, Swansea SA2 8PP,
                         UK.

Stuart L. Barr   School of Geography,
                         University of Leeds,
                         Leeds, LS2 9JT,
                         UK.

Mike Batty   Centre for Advanced Spatial Analysis,
                         University College London,
                         1–19, Torrington Place,
                         London, WC1E 6BT,
                         UK.

Yves Baudot   Laboratoire de Télédétection et d'Analyse Régionale,
                         Université Catholique de Louvain,
                         Batiment Mercator,
                         Place Louis Pasteur,
                         B1348 Louvain-la-Neuve,
                         Belgium.

Alberta Bianchin   DAEST – IUAV,
                         S. Croce 1957,
                         30135 Venezia,
                         Italy.

Pietro Antonio Brivio   Remote Sensing Department,
                             Instituto Ricerca Rischio Sismico,
                             National Research Council,
                             56 via Ampère,
                             20131 Milano,
                             Italy.

Jean-Paul Donnay   Unité de Géomatique,
                             Laboratoire SURFACES,
                             Université de Liège,
                             7 place du 20 Aout,
                             B4000 Liege,
                             Belgium.

Ben Gorte   International Institute for Aerospace Survey
                             and Earth Science (ITC),
                             Hengelosestraat 99,
                             P.O. Box 6,
                             7500 AA Enschede,
                             The Netherlands.

David Howes   NCGIA,
                             State University of New York,
                             Wilkenson Quad,
                             Buffalo,
                             NY 14261-0023,
                             USA.

Paul Longley   Centre for Advanced Spatial Analysis,
                             University College London,
                             1–19, Torrington Place,
                             London, WC1E 6BT,
                             UK.

Marc Mangolini   Aérospatiale Service Télédétection,
                             BP 99,
                             06332 Cannes la Bocca Cedex,
                             France.

Victor Mesev   School of Environmental Studies,
                             University of Ulster,
                             Cromore,
                             Coleraine, BT52 1SA,
                             UK.

Lasse Møller-Jensen   Institute of Geography,
University of Copenhagen,
Oster Volgade 10,
DK-1350, Copenhagen,
Denmark.

Martino Pesaresi   EC Joint Research Centre,
SAI/EGEO,
TP 441,
21020 Ispra (VA),
Italy.

Thierry Ranchin   Remote Sensing Group,
Centre d'Energétique,
l'Ecole des Mines de Paris,
BP207 06904,
Sophia Antipolis Cedex,
France.

David Unwin   Department of Geography,
Birkbeck College,
University of London,
7-15, Gresse Street,
London,
W1P 1PA, UK.

Lucien Wald   Remote Sensing Group,
Centre d'Energétique,
l'Ecole des Mines de Paris,
BP207 06904,
Sophia Antipolis Cedex,
France.

Christiane Weber   Laboratoire Image et Ville,
UPRES-A 7011 CNRS,
Faculté de Géographie et d'Aménagement 3,
rue de l'Argonne,
67000 Strasbourg,
France.

Eugenio Zilioli   Remote Sensing Department,
Instituto Ricerca Rischio Sismico,
National Research Council,
56 via Ampère,
20131 Milano,
Italy.

Preface

The core of this book arose from a specialist meeting on remote sensing and urban analysis, which was held in the University of Strasbourg, France. The meeting was sponsored by the European Science Foundation's (ESF) GISDATA Programme and considered four research themes in urban remote sensing, namely (i) cartographic feature extraction and map updating, (ii) delimiting urban agglomerations, (iii) characterizing urban structure, settlements and population distribution, and (iv) urban modelling.

The application of remote sensing to the study of human settlements, perhaps more than any other use of this technology, relies on the availability of ancillary data and the advanced spatial analytical functions commonly associated with, but not always present in, Geographical Information Systems (GIS). Thus, while remote sensing can help to provide us with an accurate assessment of the physical characteristics of urban areas — including their size, shape and rates of growth, as well as the principal cover types associated with their constituent land parcels — relatively little can be said about the economic activities and socio-economic conditions that prevail within them, without recourse to exogenous data sets. The demands that this places on remote sensing specialists and urban planners alike are four-fold:

1. There is a continuing requirement to develop advanced image-processing techniques which can be used to derive accurate and consistent information on the physical structure and composition of urban areas from remotely-sensed images. Work in this area is being conducted against a background of continued development in remote sensing technology, most notably through the introduction of very high spatial resolution optical satellite sensors, but also via the increasing availability of alternative sources of image data, such as SAR (Synthetic Aperture Radar) and LiDAR (Light Detection And Ranging). While these offer new possibilities, each also presages new problems. There is a danger, therefore, that remote sensing specialists will expend most of their effort addressing the opportunities and problems posed by each new technological advance, rather than seeking operational solutions to the use of existing systems.

2. There is a need to move beyond simply mapping the physical *form* of urban areas, to provide indicators of their social and economic *functioning*. This is far from straightforward, since many of the functional properties that are of interest to urban planners — such as land use and population density — cannot be observed *directly* by means of remote sensing. A more complex chain of inference is required: one that incorpo-

rates ancillary spatial data and that examines the spatial and structural properties of urban areas represented in the digital image, not just their spectral characteristics.

3. The previous point implies that new models and spatial analysis tools need to be developed to allow integrated analysis of remotely-sensed images and socio-economic data. One example of this is the need to transform socio-economic data — typically referenced by geographical units defined in terms of (vector) administrative boundaries — into raster format. Current GIS packages provide some of the necessary functionality, but not always to the level of sophistication required.

4. Finally, the communication between urban planners and remote sensing specialist must be improved, so that the former understand the full potential and limitations of remote sensing, while the latter design sensor systems and data-processing techniques that are relevant to the real, as opposed to the perceived, needs of urban planners.

This book addresses various aspects of these four points, emphasizing the methodological issues involved in the analysis of remotely-sensed data of urban areas. In many cases we have asked individual contributors to the original meeting in Strasbourg to work together to develop the chapters presented in this book. This serves two purposes: first, it forces the authors to draw out the common themes and synergies between their individual research projects, rather than relying on the reader to infer them; second, it enables them to address many of the issues that arose in discussion both during and after the GISDATA meeting. We hope that this raises the vision of the book as a whole, beyond a loose collection of papers describing separate research projects, to a more rounded assessment of the potentials and limitations of remote sensing for urban analysis.

Part I presents a set of chapters that deal with various aspects of deriving information on the physical structure and composition of urban areas from remotely-sensed data — addressing the first of the major issues outlined above.

In chapter 1 (Donnay *et al.*, this volume), we review the 'state-of-the-art' in urban remote sensing to provide some context for the rest of the book. Ranchin *et al.* (Chapter 2) then present a method for combining ('fusing') remotely-sensed images acquired at different spatial resolutions and by different satellite sensors to maximize the information available on urban areas. Their method, based on the wavelet transform, is used to merge the structural detail provided by very high spatial resolution panchromatic images with the spectral information inherent in lower resolution multispectral data sets. Since their method preserves the radiometric values of the original data, the output can be used not only for visual inspection purposes, but also for further quantitative analysis. The spatial information content of the 'raw' digital image data is also examined by Brivio and Zilioli (Chapter 3) using geostatistical techniques. More specifically, they demonstrate that the parameter values derived by fitting experimental semi-variograms to image data can be used to distinguish different types of urban environment. The final chapter in this part of the book presents the work of Pesaresi and Bianchin (Chapter 4) who explore the use of mathematical morphology operators as a means of distinguishing urban areas based on their grey-level 'texture' in high spatial resolution optical images. They demonstrate the application of this technique to satellite sensor images of a number of towns and cities in northern Italy.

Part II moves beyond a consideration of the raw radiance values reported in the remotely-sensed data to produce information on urban land cover and land use via statistical classification and structural pattern recognition.

The use of statistical classification algorithms in image processing has received extensive and detailed attention in the remote sensing literature over the last thirty years and, as a result, is now considered to be very much a 'standard' technique. Yet, despite this, the accuracy of classifications performed on images acquired over urban areas is often disappointingly low. Mesev *et al.* (Chapter 5) review the probable reasons for this and examine a number of practical techniques that can be used to improve the accuracy of per-pixel maximum-likelihood classifiers, including modifying the prior probabilities of the candidate classes on the basis of ancillary spatial information and iterative updating of the 'priors' using the posteriori probabilities.

The chapters by Bähr (Chapter 6) and Barnsley *et al.* (Chapter 7) examine a number of alternative approaches to analyzing remotely-sensed images of urban areas. Bähr, for example, considers the use of image segmentation techniques to investigate land-cover change detection within and around urban areas. The methods are informed by ancillary spatial data from the national mapping base, but are also used to explore the potential for semi-automated updating of these data by means of remote sensing. This chapter also considers the derivation of information on the three-dimensional form of urban areas using digitized aerial photography and semantic networks. Barnsley *et al.* follow on from this with a more in-depth review of structural (or syntactic) pattern recognition techniques for analyzing remotely-sensed images. Their objective is to deduce the nature of urban land use from the morphological properties of, and structural relations between, discrete land cover parcels identified in the image.

Part III examines various methods of delimit the overall extent of urban areas and of estimating their human population.

Weber (Chapter 8), for instance, focuses on the problem of producing a simple, objective definition of urban areas and urban agglomerations. The problem is of great practical interest, and Weber suggests that remote sensing has a central role to play in defining and measuring what is urban and what is not — one that can be applied consistently across different nation states. Longley and Mesev (Chapter 9) continue the broad theme of the morphology of urban areas with a consideration of their fractal dimension. They use fractal measures derived from land cover classifications of remotely-sensed images as a means of comparing the morphology of different urban areas and of understanding the processes that have underpinned their development over time. The temporal development of urban areas, as revealed in remotely-sensed images, is also explored by Batty and Howes (Chapter 10), who show how a combination of remotely-sensed and ancillary spatial data (on the age, size and taxable value of individual building lots) can be used to reconstruct the past development of a city. Perhaps more importantly, they present a convincing analysis that relates the different stages of development evident in the remotely-sensed data to changes in the dominant mode of transport (e.g.rail, car) and economic activity of the city concerned (Buffalo, NY, USA).

The final two chapters examine issues relating to the estimation of urban population density from remotely-sensed images. Donnay and Unwin (Chapter 11) are concerned with the creation of 'surface models' with which to analyze the spatial distribution of selected geographical variables, of which population density is one example. In this context, the raster image data are used to generate a continuous model of population density from discrete point- or parcel-referenced census data. Baudot (Chapter 12) uses similar spatial disaggregation techniques to estimate population counts and densities in fast-growing cities of developing countries. This work is of particular significance, given the paucity of information available via traditional means (i.e. decennial census) and its importance for strategic national and international planning purposes.

The book concludes with an analysis of the way forward in terms of the research agenda and the practical application of remote sensing for urban analysis.

Finally, we would like to take this opportunity to thank all of the participants at the meeting, not merely the authors of the chapters in this book, for their valuable contributions to the general discussion. The updated outcome of this discussion — this book — is not a complete survey on the matter; the list of the topics included here is far from exhaustive, even in terms of those covered at the Strasbourg meeting. Nevertheless, at a time when interest in the use of remote sensing for urban analysis is growing rapidly, we hope that this book will act as a catalyst for further basic research and, perhaps more importantly, the development of operational applications.

Jean-Paul Donnay, Mike Barnsley and Paul Longley,
6th October 2000

The European Science Foundation (ESF) is an association of 67 Member Organisations devoted to scientific research in 23 European countries. Since it was established in 1974, it has coordinated a wide range of pan-European scientific initiatives, and its flexible organisation structure means it can respond quickly to new developments.

The ESF's core purpose is to promote high quality science at a European level. It is committed to facilitating cooperation and collaboration in European science on behalf of its principal stakeholders (Member Organisations and Europe's scientific community). This cross-border activity combines both 'top-down' and 'bottom-up' approaches in the long-term development of science.

The work of the various parts of the ESF's organisation is governed by a series of shared values, of being Pan-European, Multidisciplinary, Flexible, Independent, Rigorous and Open. The Foundation is committed to providing scientific leadership through its networking expertise and by ensuring that there is a European added value to all of its initiatives and projects.

This volume is the ninth in a series arising from the work of the ESF's Scientific Programme on Geographic Information Systems: Data Integration and Database Design (GISDATA). This programme was launched in January 1993 and through its activities has stimulated a number of successful collaborations among GIS researchers across Europe.

Further information on ESF activities in general can be obtained from:

European Science Foundation
1 quai Lezay-Marnesia
F-67080 Strasbourg Cedex
FRANCE
Telephone +33 (0)3 88 76 71 25
Fax +33 (0)3 88 76 71 80

GISDATA Series

Series editors Ian Masser and François Salgé

PART I

INTRODUCTION

CHAPTER 1

Remote Sensing and Urban Analysis

Jean-Paul Donnay, Michael J. Barnsley and Paul A. Longley

1.1 Introduction

Aerial photography has long been employed as a tool in urban analysis (Jensen 1983, Garry 1992). Indeed, this form of remote sensing is still extensively used today and can now benefit from digital image-processing techniques, provided of course that the photographs are digitized first (Futz 1996). For reasons of their widespread availability, frequency of update and cost, however, the focus of urban remote sensing research has shifted more towards the use of digital, multispectral images, particularly those acquired by earth-orbiting satellite sensors. This trend was initiated with the advent of what might be described as 'first generation' satellite sensors, notably the Landsat MSS (Multispectral Scanning System), and was given further impetus by a number of second generation devices, such as Landsat TM (Thematic Mapper) and SPOT HRV (High Resolution Visible). Data from the former were initially used to analyze regional urban systems and for exploratory investigations of some of the larger cities in North America (Forster 1980, Jackson *et al.* 1980, Jensen 1981, Jensen 1983). The availability of still higher spatial resolution (20m/10m) images from the latter enabled more detailed studies of the older, more compact urban areas characteristic of Europe (Welch 1982, Forster 1983, Baudot 1997). The advent, over the last few years, of a third generation of very high spatial resolution (<5m) satellite sensors is likely to stimulate the development of urban remote sensing still further (National Remote Sensing Centre 1996, Aplin *et al.* 1997, Fritz 1999). The data they produce should facilitate improved discrimination of the dense and heterogeneous milieu of the old urban cores that are characteristic of European cities (Ridley *et al.* 1997), and will also help to disentangle the urban fabric in the rapidly expanding agglomerations and 'edge cities' of many developing countries.

The large number of urban remote sensing investigations that has been undertaken since the launch of the first SPOT satellite in 1986 might be taken to imply that the tool is already operational and perhaps worthy of widespread development and application through national or international programmes (Dureau and Weber 1995, Weber 1995, Dubois *et al.* 1997). In that sense, the pan-European 'Remote Sensing and Urban Statistics' initiative organized by the European statistical agency, Eurostat, is a case in point (Eurostat 1995). The subsequent work of the Commission of the European Community's Centre for Earth Observation (CEO) also deserves mention in this context (Churchill and Hubbard 1994, Galaup 1997). Despite these positive developments, however, doubts remain in some quarters about the potential for operational application of remote sensing to map and monitor urban areas (Ehlers 1995, Donnay 1999). There is, for example, a concern about their robustness and reliability. At the same time, many of the results reported in the scientific literature are of only marginal relevance to the day-to-day practices of urban planners and statisticians, who by-and-large remain unconvinced. Moreover, the case for urban remote sensing has not been helped by a degree of 'over-selling' both by some vendors of satellite sensor images and in some academic quarters.

An objective assessment of the current status and future potential of urban remote sensing is therefore required. This should include an evaluation of how the methods and results of scientific analyses — based essentially on inference and indirect identification of the form, structure and composition of urban areas — can be made more directly useful to practitioners. It should also highlight the elements of urban remote sensing that can justifiably be considered to be operational, as well as those which may become so in the near future. The need to address these issues formed the rationale for a European Science Foundation (ESF) meeting held in Strasbourg, France, and the chapters in this book have since developed as a consolidated response. Before turning to these, however, it is appropriate to appraise the 'state-of-the-art' in this field. Our review here is necessarily selective: for the most part the references are given for example only and the interested reader is invited to consult more widely the references at the end of each of the other chapters.

1.2   State-of-the-Art in Urban Remote Sensing

1.2.1   The Demand for Higher Spatial Resolution Data

It has been suggested that the single most important technical issue in urban remote sensing is the spatial resolution of the image data (Welch 1982). We have already noted that the development of urban remote sensing was stimulated by the advent of so-called 'second generation' satellite sensors with a spatial resolution of between 10m and 30m. The current sense of frustration with some of the methods of urban remote sensing can be traced to the relatively slow dissemination of data from the most recent generation of satellite sensors, which are expected to deliver image data at spatial resolutions as fine as 1m. Pending widespread availability of such data, the principal remote sensing sources are the SPOT, IRS and Landsat series of satellites, as well as data from airborne multispectral scanners and digitized aerial photography (Corbley 1996, Ridley *et al.* 1997). Thus, many analysts are already working with data of a spatial resolution finer than 5m, and are anticipating still greater discrimination of the urban fabric when the new data sources finally come on stream. To emphasize the significance of these new data sources, it is worth bearing in mind that this level of spatial resolution corresponds to scales of analysis of between 1:10,000 and 1:25,000 (ignoring the effects of relief distortion etc.) — that is to say, scales typical of projects dealing with urban planning, though not necessarily building and facility management.

The use of high or very high spatial resolution images, however, brings with it some major problems. First, the majority of the highest resolution images are presently recorded in panchromatic mode only; and, second, the mass of corresponding information creates difficulties in terms of image storage, data exchange and processing time. The solutions suggested for solving the first problem revolve around the use of various methods of data fusion; merging the higher resolution panchromatic data with lower resolution multispectral data (Jones *et al.* 1991, Ackerman 1995, Ranchin and Wald 2000, Ranchin et al. this volume). Transforms in the colour space (RGB/HSI), principal components analysis, spatial filtering and wavelet methods are among the most common means of achieving such integration (Carper *et al.* 1990, Chavez *et al.* 1991, Wald *et al.* 1997, Pohl and Van Genderen 1998). Additionally, one may observe that the fusion issue is, and will remain, of prime importance in diachronic analyses using images recorded by different generation sensors (i.e. with different spatial resolutions). The problems concerning the volume of data might appear to be of secondary importance given rapid and continuing improvements in computer technology. Nevertheless, they remain significant at the scale of the urban region (typically several tens of square kilometers) and they necessitate a consideration of data compression techniques. In this context, wavelet methods present one promising avenue of research and application. They not only allow efficient data compression while preserving the original spectral values, but can also be used to fuse images at different resolutions, thereby simultaneously dealing with the two main pitfalls outlined above.

1.2.2   Attempts to Enhance the Classification Process

Another methodological issue that must be taken into account is the land use classification process. *A priori* this problem is not restricted to urban analysis: remote sensing is fundamentally a science of inference and, even though a few Earth surface properties can be measured directly, remotely-sensed data are more generally surrogates for the actual variables of interest. In non-urban systems, it is generally possible to derive relatively straightforward, direct relationships between the spectral response of the two major components of terrestrial ecosystems (water and vegetation) and land use. By contrast, the urban milieu poses far greater problems, in that identical spectral reflectance values can correspond to very different land uses and diverse functions. The consequences for urban remote sensing investigations are two-fold: first, the quality of urban classifications is typically poorer than those obtainable for non-urban scenes (both in terms of the number of individual classes that can be recognized and the accuracy with which they can be identified); and, second, the land use typologies that can be retrieved by means of spectral 'signatures' may be different from the functional nomenclatures used in urban analysis.

To improve the results of automatic and semi-automatic classification, and to restructure the typologies to meet the needs of urban practitioners, remote sensing investigations generally augment digital image-processing with computer-assisted visual interpretation (Lenco

1997, Collet 1999). Compared to the classical procedures of digital image interpretation, however, this process is both time-consuming and is particularly subjective. Nevertheless, any enhancement of classification results, even if only a few percentage points, is usually considered to be worthwhile — much more so than in other domains where remote sensing is employed. Perhaps this also explains why developments in automated image classification techniques often find early application in the urban realm (Wilkinson 1996).

Attempts have been made to improve the results achieved using traditional per-pixel classification algorithms by using *a priori* probabilities or *a posteriori* processing (Barnsley and Barr 1996, Mesev *et al.* this volume). New algorithms, such as the so-called 'soft' classifiers (based on Bayesian probabilities, Dempster-Shafer and fuzzy set theories), have also been employed because they are able to cope more effectively with the complex spatial mixture of spectrally distinct surface materials characteristic of many urban pixels (Zang and Foody 1998). Other forms of image classification are also widely employed, including image segmentation routines (Conners *et al.* 1984, Terrettaz 1998). Ancillary (exogenous) spatial data sets, such as those pertaining to the boundaries of administrative units in towns and cities, are used to inform and constrain the classification process (Weber 1993).

Methods of combining spectral data with measures of urban form and texture are particularly numerous. Some are based on information extracted from the satellite sensor images themselves (Baraldi and Parmiggiani 1990, Forster 1993, Dawson and Parsons 1994, He *et al.* 1994, Pesaresi and Bianchin this volume), while others make use of ancillary data (Forster 1984, Sadler and Barnsley 1990, Weber 1993, Harris and Ventura 1995). The measures employed vary from comparatively simple indices of image texture (Durand *et al.* 1994) to estimates of fractal dimension (De Cola 1993, Frankhauser 1993, Longley and Mesev this volume).

Knowledge-based methods and artificial neural networks (ANNs) have been used to classify urban areas in satellite sensor images, exploiting the basic spectral data and information on the textural and contextual properties of the observed scene (Wharton 1987, Moller-Jensen 1990, Wang 1992, Halounova 1995). Still other approaches exist, notably those making use of diachrony (Nicoloyanni 1990, Gomarasca *et al.* 1993).

Generally speaking, however, these techniques are experimental and not yet included in standard image-processing software. In this respect it is worth noting that many of the functions required may be adapted from raster based GIS software. Faced with so many new and hybrid techniques, the end-users' confusion is understandable.

1.2.3   From Urban Morphology to the Human Dimension

In urban planning, the land use map generated by thematic classification of a satellite sensor image can be viewed as an end-product *per se*, as it constitutes a document showing the prevailing situation before any planning action. However, it can also be considered to be a starting point for further analyses that, *ipso facto*, will propagate the qualities and the errors inherent in the classification.

A common approach to post-classification image processing consists in the analysis of the spatial pattern of the various land use categories, especially the residential ones. This kind of morphological analysis can bring immediate practical results, such as the delineation of urban morphological agglomerations (Donnay 1994), but it also provides a means of modelling urban space filling and development. In this latter context, most of the well-known urban models have been tested, ranging from the 'old' Clark's model to new models based on the fractal theory. In this way, urban structure and land use densities can be extracted from classified and/or segmented images (De Keersmaecker 1989, Sadler *et al.* 1991, Bianchin and Pesaresi 1992, Batty and Longley 1994).

Frequently, the modelling approach makes use of a new inference process in which the land use categories are related to an urban variable, most often the human population: this approach has, of course, been widely used in the past in the analysis of aerial photography (Adeniyi 1983, Lo 1989). The simplest application involves reallocating the population figures, originally referenced by census unit, according to the spatial distribution of residential land (Langford *et al.* 1990). This can lead to an evaluation of net density figures (i.e. densities given by the ratio between the population and the area identified as being residential land) and then to a kind of visualization that is similar to dasymetric maps (Monmonier and Schnell 1984). If one further assumes that there is a close relation between different categories of residential land use on the one hand and different population densities on the other, the reallocation of the population figures may be refined still further. The same hypothesis may be used *a contrario* to estimate population figures in the absence of a formal census. This approach has been applied, via a stratified sampling procedure, to the suburbs of a number of metropolitan areas in the Third-World (Dureau 1990, Lo 1995).

The modelling of land use and/or population densities can also make use of surface models. In this case, it is a question of transforming a spatially discrete distribution into a continuous one. This is often carried out by passing a convolution kernel over the whole image. The procedure, which is similar in many respects to the application of a low-pass filter, provides a surface model of population density when applied to an image of the population figures reallocated over various residential land use categories (Langford and Unwin 1994). By modifying this procedure slightly, it can also be used to generate a potential surface of the population (Nadasdi 1995). These surfaces can then be used for further, detailed socio-economic analyses (Bracken 1991, Martin 1991, Donnay 1992, Donnay and Thomsin 1994, Thurstain-Goodwin and Unwin 2000).

1.2.4 Human Connotation of the Physical Dimension

The extraction of planimetric and 2.5-D/3-D features from remotely sensed images, whether classified or not, constitutes a very dynamic research domain in urban remote sensing. Classical photogrammetric methods, originally developed in the context of analogue and digitized aerial photographs, are increasingly being applied *mutatis mutandis* to high spatial resolution digital satellite sensor images (Ebner *et al.* 1991). The data so derived are being used in various ways. For example, information on road networks can be used for urban transport planning (Wang 1992, He *et al.* 1996, Muller and Brackman 1997, Couloigner 1998) or, more generally, in the map-updating process (Muller *et al.* 1994). Likewise, estimates of building height can be used to complete urban cadastres (Schram 1995), to assist in siting telecommunication infrastructure and, more generally, for environmental impact assessment and urban landscape ('townscape') studies (Hartl and Cheng 1995, OEEPE 1996, Ridley *et al.* 1997).

The spatial patterning of different land uses within the urban boundary, assessed and visualized by means of thematically classified satellite-sensor images, is a valuable source of information with which to analyze the urban environment. Such data can add social connotation to the mere physical infrastructure of roads and buildings, conveying information on the availability and distribution of urban 'green space' and, indirectly, on the overall quality of life (Donnay and Nadasdi 1992, Weber and Hirsch 1991, Baudot and Wilmet 1993, Richman 1999). Moreover, images of urban land use help to support and inform numerical simulations of urban growth. In this context, land use images provide the initial conditions on which forecasts can be based, as well as estimates of the areas and types of land cover that will be affected (Méaille and Wald 1990, Pathan *et al.* 1993). Finally, thermal infrared data acquired over urban areas during the day and at night can be used to monitor the so-called 'heat island' effect associated with urban areas, as well as atmospheric pollution (Nichol 1988, Eliasson 1992, Gallo *et al.* 1993, Kim 1992, Poli *et al.* 1994, Lo *et al.* 1997, Cornélis *et al.* 1998).

Clearly, the number and range of applications that can make use of information on the physical dimensions of urban areas derived from remotely-sensed images are much more extensive than the few examples discussed here. The aim has been to give a general overview and to show how the types of information that might be derived can be used to address various human and social issues pertaining to the urban environment.

1.3   Urban Remote Sensing Issues

Given the renewed interest currently being expressed by certain sections of the planning community, and in light of recent developments in sensor and computing technology, a re-evaluation of remote sensing as a tool for mapping and monitoring urban areas is now appropriate. From the viewpoint of techniques and methodologies, urban remote sensing shows an indisputable dynamism and seems to be willing and able to take advantage of each new technical and methodological innovation in the wider arena of satellite remote sensing. However, the operational potential of urban remote sensing will depend on its capacity to respond to the practical requirements of urban practitioners and on how rapidly the latter group can integrate remotely-sensed data into their everyday activities. In this context, urban remote sensing must be able to provide planners with certain, key data sets that are pertinent to urban areas, notably:

1. the location and extent of urban areas;

2. the nature and spatial distribution of different land use categories within urban areas;

3. the primary transportation networks and related infrastructure;

4. various census-related statistics and socio-economic indicators;

5. the 3-D structure of urban areas for telecommunications (inter-visibility) and Environmental Impact Assessment (EIA) studies; and

6. the ability to monitor changes in these features over time.

To fulfill these requirements, one must first resolve some ambiguities surrounding the detection, identification and analysis of urban features by means of remote sensing. This, in turn, requires a close collaboration between urban planners and remote sensing specialists.

1.3.1 Detection of Urban Features

In assessing the utility of current and future remote sensing systems against the requirements outlined above, it is important to address the following questions:

- Are the features/entities of interest to urban planning agencies intrinsically discernible by means of remote sensing, in terms of differences in their spectral, spatial, temporal, angular or polarization properties?

- What are the characteristic features of urban areas and their sub-elements that allow them to be distinguished/identified in remotely sensed data?

- Are current data-processing techniques appropriate/optimal to extract information on these features from the resultant image data?

- Can greater use be made of syntactic pattern recognition and artificial neural network techniques to derive information on urban entities, given continuing improvements in spatial resolution, particularly for new satellite sensors?

- Should such techniques now replace traditional statistical pattern recognition procedures in the analysis of remote sensed images, given the spatial heterogeneity of urban scenes and recent advances in sensor technology?

- What rôle should ancillary spatial data and Geographical Information Systems play in the image analysis/image understanding process?

A number of these questions are addressed in greater detail in the following chapters, but we will first provide some general observations here.

1.3.2 Identification of Urban Features

Finding simple, objective and repeatable criteria with which to define and identify coherent, meaningful territorial entities constitutes one of the main obstacles to an understanding of urban areas today (Pumain *et al.* 1992). Satellite remote sensing can make a vital contribution in this context, since it provides regular, repetitive data from a single, consistent source.

Identifying urban phenomena and delimiting them through space by means of remote sensing involves two key steps: first, detecting the morphological variability of settlements (e.g. city centres, suburbs, urban fringes, peripheral clusters, dilution areas or axes, conurbations); and, second, delimiting urbanized entities that are relevant to national and international statistical systems, so that they can be compared over space and time. It is important to note, however, that the delineation of urban areas cannot simply be reduced down to drawing a line around a concentrated city, itself organized in discrete, homogeneous blocks of land use. This is because urban areas frequently comprise a set of scattered settlements intertwined with other territorial elements, which only indirectly display a unified structure. The simple description of this kind of urbanization requires those elementary structures to be 'named' and 'identified', and implies that their own internal structure be studied at the higher level. Thus, structure is the key issue for analyzing and detecting this kind of urbanization.

1.3.3   Analysis of Urban Features

We have repeatedly noted that satellite remote sensing is an important source of information for urban analysis at the territorial scale. In that sense, digital image data not only provide a means of exploring and exemplifying existing hypotheses and models, but also for constructing new theories of urban areas by defining and identifying relevant spatial entities and examining the relationships between them. Despite this, the qualitative (categorical/ordinal scale) nature of the land use categories identified through the image classification process, and the discrete spatial distribution of the regions that this defines, limit the value of such images in mathematical models of urban areas. The latter frequently presuppose the presence of one or more quantitative (interval/ratio scale), spatial continuous variables. Consequently, land use data sets derived from remotely-sensed images may need to be further transformed if they are to be readily integrated in many urban models. Where this is achieved, the transformed data may then be used in a much wider range of applications, including:

- physical planning (e.g. viewshed analysis, impact assessment and other environmental issues);

- economic planning (e.g. accessibility, location analysis, and transport studies);

- social planning (e.g. population and other socio-demographic distributions, urban structures); and,

- forecasting models (e.g. diffusion and urban growth).

We note, in passing, that the accuracy and reliability of the results derived for any of the applications outlined above is dependent on the quality of the input data (e.g. the accuracy and reliability of the land use classification and the appropriateness of the nomenclature/classification scheme employed), and that the process of transforming the image data is not a neutral one (i.e. it may introduce error or bias that will vary, for example, with the scale of analysis).

1.4   Urban Remote Sensing and GIS

Although we have frequently referred to GIS in this review, it has tended to be in an indirect fashion. Nevertheless, the key rôle of GIS in providing a framework for spatial analysis of remotely-sensed data products and other sources of spatial data on urban areas is recognized. The integration of GIS and remote sensing has justifiably received widespread and extensive attention in the recent literature. Most of the scientific and commercial journals have devoted at least one special issue to this subject and this trend will likely increase over the next few years. According to Wilkinson (1996, p.85), the interface between GIS and remote sensing can be envisaged in one of three different ways:

a) remote sensing can be used as a tool to gather data sets for use in GIS;

b) GIS data sets can be used as ancillary information with which to improve the products derived from remote sensing and;

c) remote sensing data and GIS data can be used together for modelling and analysis.

Each of these views is pertinent to urban remote sensing. Urban analysis needs to exploit the capabilities of remote sensing systems in terms of their spatial coverage and detail but, because of the limitations in terms of distinguishing all aspects of the urban milieu solely on the basis of their spectral reflectance properties, classifications, analyses and models of urban areas will always be dependent on ancillary spatial information and the analytical capabilities of GIS. Ultimately, urban remote sensing is not only a meeting point for the social and physical sciences, but it is also a field of research that forms a bridge between remote sensing and GIS.

1.5 References

Ackerman, F., 1995, Sensor- and data-integration — The new challenge,, In *Integrated sensor orientation: theory, algorithms and systems*, edited by I. Colomina, and J. Navarro (Heidelberg: Wichmann), pp. 2–10.

Adeniyi, P. O., 1983, An aerial photographic method for estimating urban population. *Photogrammetric Engineering and Remote Sensing*, **49**, 545–560.

Aplin, P., Atkinson, P. M., and Curran, P. J., 1997, Fine spatial resolution satellite sensors for the next decade. *International Journal of Remote Sensing*, **18**, 3873–3882.

Baraldi, A., and Parmiggiani, F., 1990, Urban area classification by multispectral SPOT images. *IEEE Transactions on Geoscience and Remote Sensing*, **28**, 674–680.

Barnsley, M. J., and Barr, S. L., 1996, Inferring urban land use from satellite sensor images using kernel-based spatial reclassification. *Photogrammetric Engineering and Remote Sensing*, **62**, 949–958.

Batty, M., and Longley, P. A., 1994, *Fractal cities: A geometry of form and function* (London: Academic Press).

Baudot, Y., 1997, L'influence de la résolution effective des données télédétectées sur les possibilités d'analyse des milieux urbains complexes, In *Télédétection des milieux urbains et périurbains.*, edited by J.-M. Dubois, J.-P. Donnay, A. Ozer, F. Boivin, and A. Lavoie (Montréal: AUPELF-UREF), pp. 3–13.

Baudot, Y., and Wilmet, J., 1993, Vegetation monitoring in urban areas using urban remote sensing. *Sistema Terra*, **2**, 66–68.

Bianchin, A., and Pesaresi, M., 1992, Approccio strutturale all'analisi di immagine per la descrizione del territorio: una esplorazione degli strumenti di morfologia matematica, Presentation at the V Convegno Nazionale dell'Associazione Italiana Telerilevamento, Milan.

Bracken, I., 1991, A surface model approach of population for public resource allocation. *Mapping Awareness*, **5**, 35–38.

Carper, W. J., Lillesand, T. M., and Kiefer, R. W., 1990, The use of Intensity Hue Saturation transformations for merging SPOT panchromatic and multispectral image data. *Photogrammetric Engineering and Remote Sensing*, **56**, 459–467.

Chavez, P. S., Sides, S. C., and Anderson, J. A., 1991, Comparison of three different methods to merge multiresolution and multispectral data: Landsat TM and SPOT panchromatic. *Photogrammetric Engineering and Remote Sensing*, **57**, 265–303.

Churchill, P., and Hubbard, N., 1994, Centre for Earth Observation (CEO). *EARSeL Newsletter*, **20**, 18–21.

Collet, C., 1999, Vers un système d'assistance à l'interprétation d'images numériques de télédétection, In *Actes du Colloque International Télédétection et Géomatique pour la gestion des problèmes environnementaux, 67e Congrès de lACFAS*, edited by A. Bannari (Ottawa: lACFAS).

Conners, R. W., Trivedi, M. M., and Harlow, C. A., 1984, Segmentation of a high resolution urban scene using texture operators. *Computer Vision, Graphics and Image Processing*, **25**, 273.

Corbley, K. P., 1996, One-meter satellites: practical applications by spatial data users — part three. *Geo Info Systems*, **6**, 39–43.

Cornélis, B., Binard, M., and Nadasdi, I., 1998, Potentiels urbains et îlots de chaleur. *Publications de l'Association Internationale de Climatologie*, **10**, 223–229.

Couloigner, I., 1998, *Reconnaissance des formes dans des images de télédétection en milieu urbain*, Ph.D. thesis, University of Nice-Sophia Antipolis.

Dawson, B. R. P., and Parsons, A. J., 1994, Texture measures for the identification and monitoring of urban derelict land. *International Journal of Remote Sensing*, **15**, 1529–1572.

De Cola, L., 1993, Multifractals in image processing and process imaging, In *Fractals in Geography*, edited by N. S. N. Lam, and L. De Cola (Englewood Cliffs NJ: Prentice Hall), pp. 213–227.

De Keersmaecker, M. L., 1989, *Potentialités de la télédétection satellitaire pour l'étude de la structure interne des villes; applications au cas de Bruxelles*, Ph.D. thesis, Catholic University of Louvain.

Donnay, J.-P., 1992, Remotely sensed data contributes to GIS socioeconomic analysis. *GIS Europe*, **1**, 38–41.

Donnay, J.-P., 1994, Agglomérations morphologiques et fonctionnelles, l'apport de la télédétection urbaine. *Acta Geographica Lovaniensia*, **34**, 191–199.

Donnay, J.-P., 1999, Use of remote sensing information in planning, In *Geographical Information and Planning*, edited by J. Stillwell, S. Geertman, and S. Openshaw (Berlin: Springer), pp. 242–260.

Donnay, J.-P., and Nadasdi, I., 1992, Usage des données satellitaires urbaines de haute résolution en modélisation urbaine: application à l'agglomération de Maastricht. *Acta Geographica Lovaniensia*, **33**, 159–169.

Donnay, J.-P., and Thomsin, L., 1994, Urban remote sensing and statistics: prospective research and applications, In *Proceedings of the Symposium: New Tools for Spatial Analysis*, edited by M. Painlo, Eurostat (Luxembourg: Office for Official Publications of the European Communities), pp. 137–145.

Dubois, J.-M., Donnay, J.-P., Ozer, A., Boivin, F., and Lavoie, A. (editors), 1997, *Télédétection des milieux urbains et périurbains*, Collection Universités francophones — Actualités scientifiques (Montréal: AUPELF-UREF).

Durand, P., Hakdaoui, M., Chorowicz, J., Rudant, J.-P., and Simonin, A., 1994, Caractérisation des textures urbaines sur image radar VARAN par approche morphologique et statistique. Application à la ville de Luc (sud-est de la France). *International Journal of Remote Sensing*, **15**, 1064–1078.

Dureau, F., 1990, Dossier sur la télédétection urbaine à lORSTOM, Technical Report 4, Villes et citadins du Tiers-Monde.

Dureau, F., and Weber, C. (editors), 1995, *Télédétection et systèmes d'information urbains* (Paris: Anthropos).

Ebner, H., Fritsch, D., and C., H. (editors), 1991, *Digital photogrammetric systems* (Heidelberg: Wichmann).

Ehlers, M., 1995, The promise of remote sensing for land cover monitoring and modelling, In *Proceedings of the Joint European Conference on Geographical Information* (Basel: AKM Messel AG), pp. 426–432.

Eliasson, I., 1992, Infrared thermography and urban temperature patterns. *International Journal of Remote Sensing*, **13**, 869–880.

Eurostat, 1995, *Pilot Project Delimitation of Urban Agglomerations by Remote Sensing: Results and Conclusions* (Luxembourg: Office for Official Publications of the European Communities).

Forster, B. C., 1980, Urban residential ground cover using Landsat digital data. *Photogrammetric Engineering and Remote Sensing*, **46**, 547–558.

Forster, B. C., 1983, Some urban measurements from Landsat data. *Photogrammetric Engineering and Remote Sensing*, **49**, 1693–1707.

Forster, B. C., 1984, Combining ancillary and spectral data for urban applications. *International Archives of Photogrammetry and Remote Sensing*, 55–67.

Forster, B. C., 1993, Coefficient of variation as a measure of urban spatial attributes, using SPOT HRV and Landsat TM data. *International Journal of Remote Sensing*, **14**, 2403–2410.

Frankhauser, P., 1993, *La fractalité des structures urbaines*, Ph.D. thesis, UFR de Geographie, Université de Paris I.

Fritz, L. W., 1999, High resolution commercial remote sensing satellites and spatial information, http://www.isprs.org/publications/highlights/highlights0402/ fritz.html.

Futz, R. C., 1996, Integration of hardcopy and softcopy exploitation. *Geomatics Info Magazine*, **10**, 6–7.

Galaup, M., 1997, User workshops to define the requirements of town/city local government departments, Technical Report HDB/97504sg, CEO RGC9D, Space Applications Institute, Joint Research Centre, Joint Research Centre, Ispra, Italy, 2 volumes.

Gallo, K. P., McNab, A. L., Karl, T. R., Brown, J. F., Hood, J. J., and Tarpley, J. D., 1993, The use of a vegetation index for assessment of the urban heat island effect. *International Journal of Remote Sensing*, **14**, 2223–2230.

Garry, G., 1992, *L'usage des photographies aériennes*, volume 3 of *Serie Environnement et Aménagement* (Paris: Les Editions du STU).

Gomarasca, M. A., Brivio, P. A., Pagnoni, F., and Galli, A., 1993, One century of land-use changes in the metropolitan-area of Milan (Italy). *International Journal of Remote Sensing*, **14**, 211–223.

Halounova, L., 1995, Comparison of neural network and maximum likelihood classifications in an urban area, In *Sensors and Environmental Applications of Remote Sensing*, edited by J. Askne (Rotterdam: Balkema), pp. 463–468.

Harris, P. M., and Ventura, S. J., 1995, The integration of geographic data with remotely sensed imagery to improve classification in an urban area. *Photogrammetric Engineering and Remote Sensing*, **61**, 993–998.

Hartl, P., and Cheng, F., 1995, Delimiting the buildings heights in a city from the shadow on a panchromatic SPOT-image: Part 2: Test of a complete city. *International Journal of Remote Sensing*, **16**, 2829–2842.

He, D.-C., Wang, L., Baulu, T., Morin, D., and Bannari, A., 1994, Classification spectrale et texturale des données dimages SPOT en milieu urbain. *International Journal of Remote Sensing*, **15**, 2145–2153.

He, D.-C., Wang, L., and Morin, D., 1996, L'extraction du réseau routier urbain à partir d'images SPOT HRV. *International Journal of Remote Sensing*, **17**, 827–834.

Jackson, M. J., Carter, P., Smith, T. F., and Gardner, W., 1980, Urban land mapping from remotely-sensed data. *Photogrammetric Engineering and Remote Sensing*, **46**, 1041–1050.

Jensen, J. R., 1981, Urban change detection mapping using Landsat data. *The American Cartographer*, **8**, 1237–1247.

Jensen, J. R., 1983, Urban/suburban land use analysis, In *Manual of Remote Sensing, Second Edition*, edited by R. N. Colwell (Falls Church, Virginia: American Society of Photogrammetry), pp. 1571–1666.

Jones, J. G., Thomas, R. W., and Earwicker, P. G., 1991, Multiresolution analysis of remotely imagery. *International Journal of Remote Sensing*, **51**, 311–316.

Kim, H. H., 1992, Urban heat island. *International Journal of Remote Sensing*, **13**, 2319–2336.

Langford, M., and Unwin, D. J., 1994, Generating and mapping population density surfaces with a geographical information system. *The Cartographic Journal*, **31**, 21–26.

Langford, M., Unwin, D. J., and Maguire, D. J., 1990, Generating improved population density maps in an integrated GIS, In *Proceedings of EGIS'90*, EGIS Foundation (Amsterdam: EGIS Foundation), pp. 651–660.

Lenco, M., 1997, Étude par télédétection des écosystèmes urbains des grands agglomérations franaises à l'échelle du 1:25,000, In *Télédétection des milieux urbains et périurbains*, edited by J.-M. Dubois, J.-P. Donnay, A. Ozer, F. Boivin, and A. Lavoie (Montréal: AUPELF-UREF), pp. 191–206.

Lo, C. P., 1989, A raster approach to population estimation using high-altitude aerial and space photographs. *Remote Sensing of Environment*, **27**, 59–71.

Lo, C. P., 1995, Automated population and dwelling unit estimation for high resolution satellite imagery: A GIS approach. *International Journal of Remote Sensing*, **16**, 17–34.

Lo, C. P., Quattrochi, D. A., and Luvall, J. C., 1997, Application of high-resolution thermal infrared remote sensing and GIS to assess the urban heat island effect. *International Journal of Remote Sensing*, **18**, 287–304.

Martin, D. J., 1991, Understanding socioeconomic geography from the analysis of surface form, In *Proceedings of EGIS' 91* (Amsterdam: EGIS Foundation), pp. 691–699.

Méaille, R., and Wald, L., 1990, Using geographical information systems and satellite imagery within a numerical simulation of regional urban growth. *International Journal of Geographical Information Systems*, **4**, 445–456.

Moller-Jensen, L., 1990, Knowledge-based classification of an urban area using texture and context information in Landsat-TM imagery. *Photogrammetric Engineering and Remote Sensing*, **56**, 899–904.

Monmonier, M., and Schnell, G., 1984, Land use and land cover data and the mapping of population density. *International Yearbook of Cartography*, **24**, 115–121.

Muller, F., and Brackman, P., 1997, Restitution de photographies satellitaires pour la creation d'une banque de données topographiques, In *Télédétection des milieux urbains et périurbains*, edited by AUPELF-UREF (Montréal), pp. 291–298.

Muller, F., Donnay, J.-P., and Kaczynski, R., 1994, Evaluation of high resolution satellite photographs for map revision up to the scale 1:25000, In *Proceedings of ISPRS Commission IV* (Athens, GA: ISPRS), pp. 304–310.

Nadasdi, I., 1995, Inventaires biophysiques de l'occupation du sol et pratique des plans d'organisation du territoire : expériences de l'Euregion Meuse-Rhin, de la Wallonie et du Grand-Duché de Luxembourg, In *Jornadas tecnicas sobre sistemas de informacion geografica y teledeteccion espacial a la ordenacion del territorio y el medio ambiente*, Vitoria-Gasteiz, pp. 171–198.

National Remote Sensing Centre, 1996, The availability of very high resolution data: scientific, legal and commercial implications for Europe: Final Report, Technical report, European Commission (DG XII-D-4 Space).

Nichol, J. E., 1988, Visualisation of urban surface temperatures derived form satellite images. *International Journal of Remote Sensing*, **19**, 1639–1650.

Nicoloyanni, E., 1990, Un indice de changement diachronique appliqué à deux scènes Landsat MMS sur Athènes (Greece). *International Journal of Remote Sensing*, **11**, 1617–1624.

OEEPE, 1996, Survey on 3D-City Models, Organisation Européenne d'études Photogrammétriques Expérimentales (OEEPE) and Institute of Photogrammetry of the University of Bonn, http://www.ipb.uni-bonn.de/OEEPE/oeepe.html.

Pathan, S. K., Sastry, S. V. C., Dhinwa, P. S., Rao, M., L., M. K., Kumar, S., Patkar, V. N., and Phatak, V. N., 1993, Urban growth trend analysis using GIS techniques — a case study of the Bombay metropolitan region. *International Journal of Remote Sensing*, **14**, 3169–3180.

Pohl, C., and Van Genderen, J. L., 1998, Multisensor image fusion in remote sensing: concepts, methods and applications. *International Journal of Remote Sensing*, **19**, 823–854.

Poli, U., Pignatoro, A. F., Rocchi, V., and Bracco, L., 1994, Study of the heat island over the city of Rome from Landsat-TM satellite in relation with urban air pollution, In *Remote sensing — From research to operational applications in the New Europe*, edited by R. Vaughan (Berlin: Springer-Verlag), pp. 413–422.

Pumain, D., Saint-Julien, T., Cattan, N., and Rozenblat, C., 1992, Le concept statistique de la ville en Europe, In *Collection Eurostat — Th'eme 3 — Série E* (Luxembourg: Office des Publications Officielles des Communautés Européennes).

Ranchin, T., and Wald, L. (editors), 2000, *Proceedings of the Third International Conference on Fusion of Earth Data*, Ecole des Mines de Paris SEE, SEE, Sophia-Antipolis 2000, Nice., EARSeL.

Richman, A., 1999, Urban Relief, CEO Product Development and Marketing projects, http://ewse.ceo.org/anonymous/vfs.pl?id=889117&name=summary. pdf.

Ridley, H. M., Atkinson, P. M., Aplin, P., Muller, J.-P., and Dowman, I., 1997, Evaluating the potential of forthcoming commercial U.S. high-resolution satellite sensor imagery at the Ordnance Survey. *Photogrammetric Engineering and Remote Sensing*, 63, 997–1005.

Sadler, G. J., and Barnsley, M. J., 1990, Use of population density data to improve classification accuracies in remotely-sensed images of urban areas, In *Proceedings of EGIS90* (Amsterdam: EGIS Foundation), pp. 968–977.

Sadler, G. J., Barnsley, M. J., and Barr, S. L., 1991, Information extraction from remotely-sensed images for urban land analysis, In *Proceedings of the Second European GIS Conference (EGIS'91)* (Amsterdam: EGIS Foundation), pp. 955–964.

Schram, M., 1995, IRSELCAD — Remote sensing data for cadastral land consolidation. *Geomatics Info Magazine*, 9, 55–59.

Terrettaz, P., 1998, *Délimitation des agglomérations et segmentation urbaine à l'aide d'images satellitales SPOT HRV. Application aux villes de Genève, Strasbourg et Liège*, Ph.D. thesis, University of Fribourg, Switzerland.

Thurstain-Goodwin, M., and Unwin, D., 2000, Defining and delineating the central areas of towns for statistical monitoring using continuous surface representations, CASA Working Paper Series 18, University College London, http://www.casa.ucl.ac.uk/towncentres.pdf.

Wald, L., Ranchin, T., and Mangolini, M., 1997, Fusion of satellite images of different spatial resolutions: assessing the quality of resulting images. *Photogrammetric Engineering and Remote Sensing*, 63, 691–699.

Wang, F., 1992, A knowledge-based vision system for detecting land changes at urban fringes. *IEEE Transactions on Geoscience and Remote Sensing,*, 31, 136–145.

Weber, C., 1993, Traitement de l'information satellitaire et modélisation urbaine : contraintes de discrimination et reproductivité, Technical paper PNTS CNRS 90N50/0074, Institut de Géographie de l'Université Louis Pasteur, Strasbourg.

Weber, C., 1995, *Images satellitaires et milieu urbain* (Paris: Hérmes).

Weber, C., and Hirsch, J., 1991, Some urban measurements from SPOT data: urban life quality indices. *International Journal of Remote Sensing*, **13**, 3251–3261.

Welch, R., 1982, Spatial resolution requirements for urban studies. *International Journal of Remote Sensing*, **3**, 139–146.

Wharton, S. W., 1987, A spectral-knowledge-based approach for urban land-cover discrimination. *IEEE Transactions on Geoscience and Remote Sensing*, **25**, 272–282.

Wilkinson, G. G., 1996, A review of current issues in the integration of GIS and remote sensing data. *International Journal of Geographical Information Systems*, **10**, 85–101.

Zang, J., and Foody, G. M., 1998, A fuzzy classification of sub-urban land cover from remotely sensed imagery. *International Journal of Remote Sensing*, **19**, 2721–2238.

PART II

THE PHYSICAL STRUCTURE AND COMPOSITION OF URBAN AREAS

Improving the Spatial Resolution of Remotely-Sensed Images by Means of Sensor Fusion: A General Solution Using the ARSIS Method

Thierry Ranchin, Lucien Wald and Marc Mangolini

2.1 Introduction

A wide range of Earth observation satellites appropriate to monitoring urban areas now exists. Each produces data at a different spatial resolution and in different spectral wavebands (see Donnay *et al.*, Chapter 1, this volume). For example, the familiar Thematic Mapper sensors on board the Landsat-series of satellites generate data at a spatial resolution of $30m$ in six spectral wavebands spanning visible blue to shortwave-infrared wavelengths, and at $120m$ in a further spectral waveband located in the thermal infrared. The HRV (High Resolution Visible) sensors on board the SPOT-series of satellites, on the other hand, provide a single-channel panchromatic (P) image at $10m$ resolution and three multispectral images (XS1, XS2 and XS3) at $20m$ resolution. Likewise, the Indian Remote Sensing satellite, IRS-1C, generates a panchromatic image at $5.8m$, three multispectral images at $23.5m$, a short-wave infrared image at $70.5m$ (LISS-3 instrument), and two further images at $188m$ (WiFS instrument).

This diversity of Earth observation sensors continues to increase. In recent years, for example, the SPOT-series of satellites has been extended, with the launch of SPOT-4. This satellite carries an enhanced version of the HRV sensor, known as HRVIR (High Resolution Visible Infra-Red), which provides data in a spectral bands equivalent to the XS of the HRV instruments, plus a new band located in the shortwave infra-red. A second sensor, known as VEGETATION, has been added to the same platform. This produces images in the same spectral bands as HRVIR, but at a spatial resolution of about $1km$. It is anticipated that the future HRVIR instruments, which will be mounted on the next satellite in this series (SPOT-5), will offer a $5m$ spatial resolution panchromatic band and $10m$ multispectral bands. By the same token, the Enhanced Thematic Mapper (ETM), which was launched aboard the Landsat 7 satellite in 1999, delivers a $15m$ panchromatic image in addition to the familiar set of $30m$ spatial resolution multispectral images. Several new civilian missions are scheduled for launch, each of which will carry a sensor capable of delivering images at a spatial resolution of up to $1m$–$2m$ in panchromatic mode and $4m$–$8m$ in multispectral mode (see also Donnay *et al.*, Chapter 1, this volume).

The technological developments described above have provided a fresh impetus for the application of satellite remote sensing to the study of urban areas. Previously, the relatively coarse spatial resolution of satellite sensors meant that remote sensing of urban areas was primarily realized through the use of airborne devices. Only very large cities with wide streets and comparatively simple geometrical structures could be studied effectively using satellite sensors. However, the continuing improvements in the spatial resolution of space-borne sensors, outlined above, mean that this situation is now changing. The challenge now is to find ways to derive the maximum value from the data that these new sensors provide. This requires the combined use of (i) the high spatial resolution panchromatic data to provide an accurate description of the size, shape and spatial structure of the principal objects/entities found in towns and cities and (ii) the slightly coarser spatial resolution multispectral data if different types of urban land cover/land use are to be classified. This operation is sometimes referred to as *data fusion* or *sensor fusion*.

Numerous approaches to sensor fusion have been proposed in the literature, though few can be considered satisfactory in terms of preserving the (spectral) information content of original images (Mangolini *et al.* 1995). In view of this, we have developed a new method of sensor fusion which 'improves' the spatial resolution of the images up to the best available in the combined data set, while at the same time preserving the spectral content of original multispectral images. Our method, known as ARSIS (Amelioration de la Résolution Spatiale par Injection de Structures), is designed to be widely and generally applicable and makes use of the wavelet transform. This Chapter provides a brief presentation of the wavelet transform (Section 2.2) and multi-resolution data analysis (Section 2.3), before introducing the ARSIS method (Section 2.4) and two examples of its application (Section 2.5). The Chapter concludes with a discussion of the principal benefits of the ARSIS method.

2.2 The Wavelet Transform

Since its early days, the field of remote sensing has been concerned with the analysis, understanding and characterization of various phenomena on the Earth surface. Remotely-sensed images provide a physical representation of these phenomena which, in most instances are non-stationary, of finite energy, and have characteristic spatial and temporal scales. Numerous mathematical tools have been developed (or adapted) and applied to remotely-sensed images in order to represent these phenomena in an effective and comprehensive way. One of the most widely-used of these mathematical tools is the Fourier transform, which attempts a 'decomposition' of the image on the basis of a set of sine and cosine functions. This results in a new representation of the phenomena under investigation, expressed in terms of spatial frequencies. It is very effective in this context, due to the precise localization of the elementary functions in the *frequency domain*. Unfortunately, these functions are not localized in the *spatial domain*. To understand the implication of this observation, consider the following (musical) example. A musical signal analyzed by means of a Fourier transformation presents all of the tonal frequencies in the original signal. It is not at all easy, however, to find the relative position of each note in the musical score from this representation. In other words, the Fourier transform gives a *global* representation of the signals or images analyzed from which it is difficult to draw conclusions or to make inferences about local effects. Another drawback of the Fourier transform is that it is designed to analyze periodic functions. Unfortunately, in the case of remotely sensed images, the phenomena rarely exhibit a simple periodic spatial structure. Hence, in such instances, the mathematical assumptions underpinning the Fourier transform are not met.

In order to analyze non-periodic signals, notably in the context of time-series data, various other transformations have been proposed, including the window Fourier transform, the cosine transform, the Gaussian transform, the Wigner-Ville transform and the wavelet transform. One of these, the wavelet transform, received renewed attention in the early 1980s in the context of seismic geology (Grossmann and Morlet 1984). Originally developed in the fields of quantum physics, signal processing and mathematics, wavelet theory was unified from a mathematical point-of-view by Meyer (1990). Closely related to multi-resolution analysis (Mallat 1989), an important property of the wavelet transform is that the window of analysis is adapted locally to the phenomena under investigation, such that it is able to provide information on local signals. If one goes back to the music example, based on the wavelet transform of the original score it is possible to find the contribution of a single note to one frequency and, at the same time, to derive information about the position of the note in the musical score. The wavelet transform thus leads to a *time-frequency* representation of the data under investigation: in the case of images, the wavelet transform leads to a so-called *scale-space* representation.

Similar to the Fourier transform, the wavelet transform enables the input signal to be decomposed on the basis of a series of elementary functions — known as *wavelets*. These are generated by dilations and translations of a single function, known as the 'mother' wavelet (Equation 2.1).

$$\Psi_{a,b} = |a|^{\frac{-1}{2}} \Psi\left(\frac{x-b}{a}\right) \tag{2.1}$$

where

a is the dilation step,

b the translation step

$a, b \in \Re$ and

$a \neq 0$.

Many such mother wavelets exist. They are all oscillating functions that are well localized in both time and frequency. All wavelets have common properties such as regularity, oscillation and localization, and satisfy an admissibility condition. For further details about the properties of the wavelets, the reader is referred to Meyer (1990) or Daubechies (1992). Despite sharing common properties, each wavelet transform leads to a unique decomposition of the signal depending on which mother wavelet is selected. In the one-dimensional case, the continuous wavelet transform of a function $f(x)$ is given by Equation 2.2.

$$WT_f(a, b) = \langle f, \Psi_{a,b} \rangle = \frac{1}{\sqrt{|a|}} \int_{-\infty}^{+\infty} f(x) \overline{\Psi\left(\frac{x-b}{a}\right)} dx \qquad (2.2)$$

where

$\Psi_{a,b}$ is defined as in Equation 2.1

$\overline{\Psi\left(\frac{x-b}{a}\right)}$ is the complex conjugate of Ψ and

$WT_f(a, b)$ represents the information content of $f(x)$ at scale a and location b. For fixed a and b, $WT_f(a, b)$ is called the *wavelet coefficient*.

The computation of the wavelet transform for each scale and each location of a signal provides a local representation of this signal. The process can be inverted so that the original signal can be reconstructed exactly (i.e. without loss) from the wavelet coefficients using Equation 2.3.

$$f(x) = \frac{1}{C_\Psi} \int_{-\infty}^{+\infty} \int_{-\infty}^{+\infty} WT_f(a, b) \Psi_{a,b}(x) \frac{da\, db}{a^2} \qquad (2.3)$$

where

C_Ψ is the admissibility condition of the mother wavelet.

Equation 2.3 can be interpreted in two ways:

• $f(x)$ can be reconstructed exactly if one knows its wavelet transform,

• $f(x)$ is a super-imposition of wavelets.

Each leads to a different application of the wavelet transform: in the first instance, to the processing of the signals; in the second, to their analysis. Discrete versions of the wavelet transform exist and are applied to signals using filters.

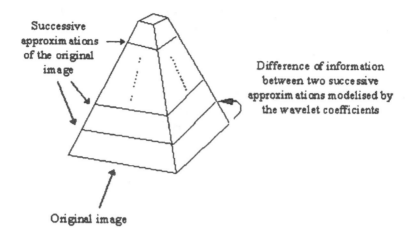

Successive approximations of the original image →

Difference of information between two successive approximations modelised by the wavelet coefficients

Original image

Figure 2.1: Pyramid representing the use of the wavelet transform for multi-resolution image analysis.

2.3   Multi-Resolution Analysis

The association between the wavelet transform and multi-resolution analysis provides a powerful and comprehensive means of processing and analyzing remotely-sensed images. The concept of multi-resolution analysis, introduced by Mallat (1989), derives from concept of Laplacian pyramids (Burt and Adelson 1983). In this approach, the size of a pixel is defined as a resolution of reference to allow a measure of local variation in the image. Note that resolution is inversely related to the scale used by cartographers. Hence, the larger the resolution of an image, the smaller the size (or the characteristic length or scale) of the smallest visible object.

Figure 2.1 provides a convenient graphical representation of multi-resolution analysis and, more generally, of pyramidal algorithms. At the base of the pyramid is the original image. Each successive level up the pyramid is an approximation of the original image and is computed from it. On climbing the pyramid, successive approximations have an ever coarser spatial resolution. The theoretical limit of multi-resolution analysis is reached when one pixel is used to represent the arithmetic mean of all of the pixels in the original image; however, due to physical constraints this limit is never reached. The base of the pyramid can also be considered to be an approximation of the landscape, as measured by the sensor.

In the context of multi-resolution analysis, the wavelet transform permits differences between two successive approximations of the same image (i.e. two successive levels of the pyramid) to be described by means of the wavelet coefficients. By inverting this process, it follows that the original image can be reconstructed without loss or error from one approximation (i.e. data at one level of the pyramid) using the wavelet coefficients. Thus, the wavelet transform and its coefficients enable us to construct a hierarchical description of the information contained in the image. Meyer (1993, p.45) uses an example drawn from the field of cartography to illustrate the use of pyramidal algorithms:

Table 2.1: Decomposition of an image by means of wavelet transformation in multi-resolution analysis.

Context image (Approximation of the original image)	Horizontal wavelet coefficient image
Vertical wavelet coefficient image	Diagonal wavelet coefficient image

'We can see from this example the fundamental idea of representing the image by a tree. In the cartographic case, the trunk would be the map of the world. By travelling toward the branches, the twigs and the leaves, we reach successive maps that cover smaller regions and give more details, details that do not appear at lower levels. To interpret the cartographic representation using the pyramid algorithm, it will be necessary to reverse the roles of top and bottom, since the pyramid algorithm progresses from 'fine to coarse'.'

The mathematical foundations of multi-resolution analysis are described in Mallat (1989). From a practical point-of-view, the computation of the approximations is performed by applying low-pass filters to the image data, while the difference of information between two successive approximations is computed using band-pass filters. Note that it is a directional algorithm and that it provides a decomposition of the original image using the scheme presented in Table 2.1. In this scheme, the context image represents the approximation to the original image. The wavelet coefficient images represent the difference of information between the original image and the context image in the diagonal, horizontal and vertical directions. The wavelet coefficient and context images are obtained by filtering and sub-sampling the original image. In the most commonly-used algorithm — the Mallat algorithm — sub-sampling is by a factor of 2. Hence, if the original image has a spatial resolution of $10m$, the first context image will have a spatial resolution of $20m$, while the wavelet coefficient images will represent the information existing between $10m$ and $20m$. The process can be iterated. From the first context image, a new context image and new wavelet coefficient images can be computed. The second context image will have the spatial resolution of $40m$, while the corresponding wavelet coefficients images will represent the information between $20m$ and $40m$. Other algorithms exist with a ratio of scales equal to $\sqrt{2}$ or $\frac{q-1}{q}$. Further details on the mathematics of wavelets and multi-resolution analysis can be found in Meyer (1990), Chui (1992a), Chui (1992b), Daubechies (1992) and Meyer (1993). A good review of wavelets can also be found in Rioul and Vetterli (1991).

In terms of the analysis and processing of remotely-sensed images, wavelets and multi-resolution analysis have been used to assess the characteristic scales of variation in geology (Besnus *et al.* 1993), in urban areas (Ranchin and Wald 1993b), and in oceanography (Ranchin and Wald 1993a). These tools have also been used to reduce the speckle in the Synthetic Aperture Radar (SAR) images (Cauneau and Ranchin 1992, Proenca *et al.* 1992, Ranchin and Cauneau 1994, Eldridge and Lasserre 1995). The wavelet transform has also been examined as a means of restoring blurred images acquired by airborne sensors

(Bruneau *et al.* 1991), of image data compression (Antonini *et al.* 1992), and of automatic image registration (Djamdji *et al.* 1993). Further applications of the wavelet transform include the analysis of the spatial and temporal structures in satellite-derived irradiance fields (Beyer *et al.* 1995), characterization of the local roughness of Digital Elevation Models (Dactu *et al.* 1996) and the analysis of surface properties of sea ice (Lindsay *et al.* 1996). Of more direct relevance to the present study, wavelet transforms have also been applied to the fusion of images and raster-maps of different spatial resolutions (Wald and Ranchin 1995), to the merging of different remotely-sensed images for updating maps of urban areas (Ranchin and Wald 1996b), and to the merging of SPOT and SAR images (Mangolini *et al.* 1993). These latter studies make use of the ARSIS method (Mangolini *et al.* 1992), which is presented in detail in the following section.

2.4 The ARSIS Method

There are four prerequisites to the sensor fusion/data merging process:

1. the images to be merged should have different spatial and spectral resolutions,

2. the images to be merged should represent the same area,

3. images should be accurately registered, and

4. no major changes should have occurred in the study area in the interval between acquisition of first and last source images.

Strictly speaking, it is possible for the last requirement not to be satisfied, in which case the aim of the data fusion process might be to perform change detection on the observed area (Ranchin and Wald 1996b). It is also important to note that the four requirements outlined above do not limit the data fusion process to images acquired by different sensors mounted on the same platform. The process can be applied to images acquired by, for example, airborne and spaceborne sensors.

Numerous methods have been developed to enhance the spatial resolution of remotely sensed images by taking advantage of the presence of one or more images of the same scene acquired by a finer spatial resolution sensor (Carper *et al.* 1990, Chavez *et al.* 1991). However, if the objective is to transform each of the candidate images to the spatial resolution of the finest resolution image in the set while retaining their spectral information content, the list becomes much shorter. Mangolini *et al.* (1995) have performed a detailed comparison of the data fusion techniques that meet these criteria, including the ARSIS method, to which the interested reader is referred. For the purpose of this Chapter, however, we will focus exclusively on the ARSIS method.

The ARSIS method was initially designed to handle images acquired in panchromatic and multispectral mode by the SPOT-HRV sensors, but has subsequently been generalized to allow the fusion of images from any two sensors having different spatial and spectral resolution. As has already been noted, the ARSIS method uses the wavelet transform and multi-resolution analysis to decompose the two images to be merged (Figure 2.2). The method, in outline, is as follows. Multi-resolution analysis using the wavelet transform is

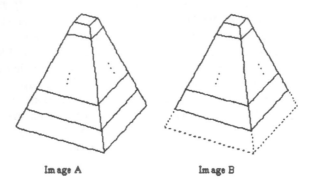

Image A   Image B

Figure 2.2: The use of the multi-resolution analysis in the ARSIS method.

applied to both image A (higher spatial resolution) and image B (lower spatial resolution), and is used to describe each image at different spatial resolutions, as well as the differences in their information content between successive approximations to the original images. The wavelet coefficients determined from multi-resolution analysis of image A — for the change of scale (or resolution) between that of image A and that of image B — are used to describe the spatial information absent from image B. This, in turn, is used to synthesize image B at the spatial resolution of image A. This can be envisaged as the process of shifting the wavelet coefficients from pyramid A to pyramid B in Figure 2.2.

It should be evident from this discussion that, if the wavelet coefficients provided by image A are used without modification, the *synthesized* image B will not be exactly equivalent to what would have been seen by sensor B had it had the spatial resolution of sensor A, This is a result of differences in, among other things, the spectral resolution and wavebands of the two images. To improve the quality of the synthesized image B still further, a model would be required which takes into account the physics of the environment so that the wavelet coefficients derived from image A can be transformed into the equivalent values for image B. Whatever form this model takes, the ARSIS method preserves the spectral content of original multispectral image, while enhancing its apparent spatial resolution. Moreover, as a validation mechanism or 'checksum', multi-resolution analysis applied to the *synthesized* image B can be used to generate an image similar to the original image B.

Figure 2.3 presents an example application of the ARSIS method to images acquired by the SPOT-HRV sensors. The example data set comprises a $10m$ spatial resolution panchromatic image and three $20m$ multispectral images (XS1, XS2 and XS3). The objective is to use the ARSIS method to reproduce the three XS images at a spatial resolution of $10m$ while preserving their original spectral content. Two iterations of multi-resolution analysis using the wavelet transform are applied to the original panchromatic (P) image and one to each of the original XS images. The mother wavelet used in this instance is the one proposed by Daubechies (1988). It is an orthogonal wavelet which permits the de-correlation of the structures between the different approximations and is implemented through a four-coefficient filter.

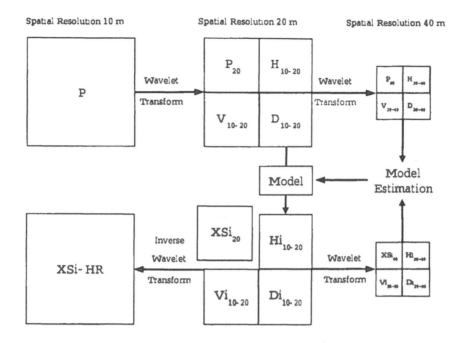

Figure 2.3: Application of the ARSIS method to SPOT-HRV P and XS images.

A model of the transformation from the P wavelet-coefficient images to the XS wavelet-coefficient images is estimated for each direction at a spatial resolution of $40m$. Various types of model can be used in this context. The simplest is the identity model; a more convenient model is one that adjusts the mean and the variance of the histogram of the wavelet-coefficient images. Whichever model is used, it should take into account the physics of both images and the correlation between them. Several such models have been tested by Mangolini *et al.* (1992). Optimum results were achieved using a model taking into account the local variation between the P and XS wavelet-coefficient images. The estimated model is then inferred at the spatial resolution of 20m. The model is subsequently used to transform the P wavelet-coefficient images representing the information lost between 10m and 20m into the corresponding XS wavelet-coefficient images. At this stage, the multi-resolution analysis is inverted, such that the XS images at 10m spatial resolution (referred to as XS*i*-HR images) are synthesized from the original XS images and the wavelet-coefficient images computed from the model.

The preceding discussion is summarized in Figure 2.4, which presents the general scheme of the ARSIS method. First, multi-resolution analysis based on the wavelet transform is used to compute the wavelet coefficients and the approximations of image A (Stage 1). The same operation is applied to the image B (Stage 2). The wavelet coefficients provided by each decomposition are used to compute a transformation model of the known wavelet-coefficients for image A to the known wavelet-coefficients for image B. This model takes into account

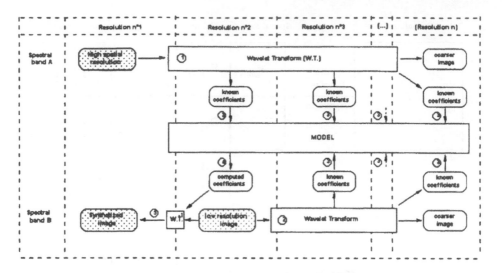

Figure 2.4: General scheme of the ARSIS method.

the physics of both images and the correlation existing between the two sets of wavelet-coefficient images (Stage 3). The model can have various forms and take into account more than one scale. It is used to compute the missing wavelet coefficients (Stage 4). Finally, by inverting the multi-resolution analysis (WT^{-1}), it is possible to synthesize image B at the spatial resolution of image A (Stage 5).

The ARSIS method has been successfully applied to merge SPOT-HRV XS and P images (Ranchin *et al.* 1994), to combine $120m$ spatial resolution data from Landsat Thematic Mapper band 6 with the $30m$ images from the other six spectral wavebands bands of this instrument (Ranchin 1993), to fuse the $30m$ Landsat Thematic Mapper bands with a $10m$ SPOT P image (Blanc *et al.* 1996), and to merge SPOT XS images with data from the $2m$ Russian panchromatic sensor KVR-1000 (Ranchin *et al.* 1996). Mangolini *et al.* (1995) proposed a method to evaluate quantitatively the different sensor fusion techniques. They show that the ARSIS method gives the best results in terms of preservation of the spectral quality. In the next section, a further two examples of the application of the ARSIS method are presented.

2.5   Examples

The first example concerns SPOT-HRV images. The images were acquired on September 11, 1990 over Barcelona (Spain), a large city located by the Mediterranean in north-east Spain. The study area covers the harbour, airport, various urban areas with roads and motorways, rivers, agricultural lots, mountainous areas and semi-natural Mediterranean vegetation. The data comprise a panchromatic image ($10m$ spatial resolution) and three multispectral images (XS1, XS2 and XS3 at $20m$ spatial resolution). These images were acquired simultaneously and have been spatially co-registered.

Figure 2.5: Original XS1 image (20*m*) covering part of Barcelona, Spain. © CNES-SPOT Image, 1990.

Figure 2.6: Synthesized XS1-HR image (10*m*) of part of Barcelona, Spain. See Figure 2.5 for a comparison with the original multispectral data.

Figure 2.7: Original SPOT-HRV XS images of part of Riyadh, Saudi Arabia. © CNES-SPOT Image, 1993.

The ARSIS method has been used to synthesize the multispectral images at a spatial resolution of 10m. The resulting data sets are referred to as XSi-HR images. Figure 2.5 presents an extract of the XS1 image at the original spatial resolution of 20m. This extract contains agricultural lots mixed with urban areas. The roads and the motorways are clearly visible, although the interchanges on the motorways are difficult to distinguish. A comparison can be made between the XS1 image (Figures 2.5) and the XS1-HR image (Figure 2.6). The original XS1 image was interpolated using a nearest-neighbour algorithm. The enhanced visual quality of the XS1-HR image is due to the injection of information from the panchromatic image, modelled in such a way as to preserve its spectral properties. Thus, in the XS1-HR image, one can clearly distinguish the interchanges on the motorways, the road network in the upper left corner of the image and several large buildings.

In the second example, the set of images comprises a SPOT-HRV XS scene of the town of Riyadh, Saudi Arabia acquired on May 16, 1993 and a Russian KVR-1000 image of the same area acquired on September 7, 1992. The three multispectral images have a spatial resolution of 20m, while the KVR-1000 image has a spatial resolution of 2m and a spectral range of 0.51μm–0.71μm. Figure 2.7 presents the original XS images, while Figure 2.8 shows the XSi-HR images at synthesized at a spatial resolution of 2m. In this case, the factor of 10 difference between the spatial resolution of the KVR-1000 and XS images is important. This area is made up of a large interchange between two urban motorways, numerous buildings, and some sandy areas. The large object on the right-hand side of the motorway, in the lower part of Figure 2.8, is a shopping mall. Due to the fine detail which appears in the synthesized image, it is possible to distinguish the structure of this mall and of the other buildings in this area. By preserving the spectral content of original XS images, the ARSIS method allows the roads and the buildings to be extracted by means of a

Figure 2.8: Synthesized XS-HR image (2*m*) covering part of the city of Riyadh, Saudi Arabia. See Figure 2.7 for a comparison with the original multispectral data.

standard multispectral classification algorithm or, indeed, any other method that requires the multispectral content provided the original images. This approach has been demonstrated successfully by Ranchin and Wald (1997), who show the improvement in the accuracy with which roads in urban areas can be extracted from remotely-sensed images using the ARSIS method.

2.6   Conclusions

This chapter has reviewed the ARSIS method of sensor fusion, which is based on multi-resolution analysis and the wavelet transform. By definition, ARSIS preserves the spectral content of original multispectral images while improving their apparent spatial resolution. The synthesized images can therefore be used for purposes other than simply visual interpretation. Several examples of the application of this method have been presented. Elsewhere, it has been demonstrated that the accuracy and the quality with which the road network can be extracted from SPOT-HRV multispectral images can be increased through the use of images synthesized by means of the ARSIS method (Ranchin and Wald 1997). The benefits of using these synthesized images for urban mapping are discussed by Ranchin and Wald (1996a). Here, the ARSIS method has been demonstrated for two sets of images covering different urban areas. In the first instance, the method was applied to SPOT-HRV images (P and XS) of Barcelona, Spain. In the second, to the merging of a SPOT-HRV XS image with 2*m* data from the KVR-1000 sensor. In other studies, we have demonstrated the general applicability of this method through its application to the fusion of Landsat Thematic Mapper (TM) channel 6 (120*m*) with the other bands of the Landsat TM sensor, of Landsat TM and

SPOT-HRV panchromatic images. Further studies have shown its value to data from sensors on board the SPOT 4 (Ranchin and Wald 1996c) and the SPOT 5 (Ranchin and Wald 1996a) missions. The next step in this work is to demonstrate the resultant improvements in the accuracy of various standard image-processing techniques applied to the synthesized multispectral data.

2.7   Acknowledgements

The authors acknowledge fruitful discussions with Michel Barlaud, Pierre Mathieu and Eric Sere. This work was partly supported by the French Space Agency (CNES) and the French Ministry of Defence (DGA).

2.8   References

Antonini, M., Barlaud, M., Mathieu, P., and Daubechies, I., 1992, Image coding using the wavelet transform. *IEEE Transactions on Image Processing*, **1**, 205–220.

Besnus, Y., Pion, J. C., Raffy, M., Ramstein, G., and Yésou, H., 1993, Traitements d'image appliqués la cartographie des formations dunaires de l'erg Akchar (Mauritanie), apport des données Landsat Thematic Mapper et de l'analyse par ondelettes d'une image SPOT panchromatique, In *Actes des troisièmes journées du réseau de télédétection de l'UREF: Outils micro-informatiques et télédétection de l'évolution des milieux* (Sainte-Foy, Québec: Presses de l'Université du Québec), pp. 253–269.

Beyer, H. G., Costanzo, C., and Reise, C., 1995, Multiresolution analysis of satellite-derived irradiance maps. An evaluation of a new tool for the spatial characterization of hourly irradiance fields. *Solar Energy*, **55**, 9–20.

Blanc, P., Ranchin, T., Blu, T., Wald, L., and Aloisi, R., 1996, Using iterated rational filter banks within the ARSIS method for producing 10m Landsat multispectral images, In *Proceedings of the conference on Fusion of Earth data: merging point measurements, raster maps and remotely sensed images*, edited by T. Ranchin, and L. Wald, SEE/URISCA (Nice: SEE/URISCA), pp. 69–74.

Bruneau, J. M., Blanc-Féraud, L., and Barlaud, M., 1991, Opérateurs de régularisation en restauration d'image: calculs et comparaisons, In *Actes de treizième colloque du GRETSI*, Juan-Les-Pins, pp. 781–784.

Burt, P. J., and Adelson, E. H., 1983, The Laplacian pyramid as a compact image code. *IEEE Transactions on Communications*, **31**, 532–540.

Carper, W. J., Lillesand, T. M., and Kiefer, R. W., 1990, The use of Intensity Hue Saturation transformations for merging SPOT panchromatic and multispectral image data. *Photogrammetric Engineering and Remote Sensing*, **56**, 459–467.

Cauneau, F., and Ranchin, T., 1992, Speckle removal in SAR images using the wavelet transform, In *Proceedings of the 12th Symposium of EARSeL: Remote Sensing for Monitoring the Changing Environment of Europe*, edited by P. Winkler, EARSeL (Rotterdam: A.A. Balkema), pp. 97–104.

Chavez, P. S., Sides, S. C., and Anderson, J. A., 1991, Comparison of three different methods to merge multiresolution and multispectral data: Landsat TM and SPOT panchromatic. *Photogrammetric Engineering and Remote Sensing*, 57, 265–303.

Chui, C. (editor), 1992a, *An Introduction to Wavelets*, volume 1 of *Wavelet Analysis and its Application* (Boston: Academic Press).

Chui, C. (editor), 1992b, *Wavelets: A Tutorial in Theory and Applications*, volume 2 of *Wavelet Analysis and its Application* (Boston: Academic Press).

Dactu, M., Dragos, L., and Seidel, K., 1996, Wavelet-based digital elevation model analysis, In *Proceedings of the 16th Symposium of EARSeL: Integrated Applications for Risk Assessment and Disaster Prevention for the Mediterranean*, EARSeL (Rotterdam: A.A. Balkema).

Daubechies, I., 1988, Orthonormal bases of compactly supported wavelets. *Communications on Pure and Applied Mathematics*, 41, 906–909.

Daubechies, I., 1992, *Ten Lectures on Wavelets*, volume 61 of *CMBS-NSF Regional Conference Series in Applied Mathematics* (Philadelphia: Society for Industrial and Applied Mathematics (S.I.A.M.)).

Djamdji, J. P., Bijaoui, A., and Manière, R., 1993, Geometrical registration of images: the multiresolution approach. *Photogrammetric Engineering and Remote Sensing*, 59, 645–653.

Eldridge, N. R., and Lasserre, M., 1995, Wavelet transform analysis and speckle filtering of Synthetic Aperture Radar (SAR) sea ice imagery, In *Proceedings of the 17th Canadian Symposium on Remote Sensing* (Canadian Remote Sensing Society), pp. 161–167.

Grossmann, A., and Morlet, J., 1984, Decomposition of Hardy functions into square integrable wavelets of constant shape. *SIAM Journal of Mathematic Analysis*, 15, 723–736.

Lindsay, R. W., Percival, D. B., and Rothrock, D. A., 1996, The discrete wavelet transform and the scale analysis of the surface properties of sea ice. *IEEE Transactions on Geoscience and Remote Sensing*, 34, 771–787.

Mallat, S. G., 1989, A theory for multiresolution signal decomposition: the wavelet representation. *IEEE Transactions on Pattern Analysis and Machine Intelligence*, 11, 674–693.

Mangolini, M., Ranchin, T., and Wald, L., 1992, Procédé et dispositif pour l'amélioration de la résolution spatiale d'images à partir d'autres images de meilleure résolution spatiale, Patent No. 92-13961.

Mangolini, M., Ranchin, T., and Wald, L., 1993, Fusion d'images SPOT multispectrales (XS) et panchromatique (P), et d'images radar, In *De l'Optique au Radar, les Applications de SPOT et ERS* (Toulouse: Cépaduès-Editions), pp. 199–209.

Mangolini, M., Ranchin, T., and Wald, L., 1995, Évaluation de la qualité des images multispectrales à haute résolution spatiale dérivées de SPOT. *Bulletin de la Société de Photogrammétrie et Télédétection*, 137, 24–29.

Meyer, Y., 1990, *Ondelettes et Opérateurs 1: Ondelettes* (Paris: Hermann).

Meyer, Y., 1993, *Ondelettes et Algorithmes Concurrents (Wavelets : Algorithms and Applications)*, Translated and revised by Robert D. Ryan (Philadelphia: Society for Industrial and Applied Mathematics).

Proenca, M. C., Rudant, J. P., and Flouzat, G., 1992, Using wavelets to get SAR images freeöf speckle, In *Proceedings of the 12th Symposium IGARSS'92, International Space Year: Space Remote Sensing* (IGARSS), pp. 887–889.

Ranchin, T., 1993, *Applications de la transformée en ondelettes et de l'analyse multirésolution au traitement des images de télédétection*, Ph.D. thesis, University of Nice-Sophia Antipolis.

Ranchin, T., and Cauneau, F., 1994, Speckle reduction in Synthetic Aperture Radar imagery using wavelets, In *Mathematical Imaging: Wavelet Applications in Signal and Image Processing, Proceedings of SPIE International Symposium on Optics, Imaging and Instruments* (SPIE), pp. 432–411.

Ranchin, T., and Wald, L., 1993a, Applications of wavelet transform in remote sensing processing, In *Proceedings of the 12th Symposium of EARSeL: Remote Sensing for Monitoring the Changing Environment of Europe*, EARSeL (Rotterdam: A.A. Balkema), pp. 261–266.

Ranchin, T., and Wald, L., 1993b, The wavelet transform for the analysis of remotely sensed images. *International Journal of Remote Sensing*, 14, 615–619.

Ranchin, T., and Wald, L., 1996a, Benefits of fusion of high spatial and spectral resolution images for urban mapping, In *Proceedings of the 26th International Symposium on Remote Sensing of the Environment and the 18th Annual Symposium of the Canadian Remote Sensing Society* (Canadian Remote Sensing Society), pp. 262–265.

Ranchin, T., and Wald, L., 1996b, Merging SPOT-P and KVR-1000 for updating urban maps, In *Proceedings of the 26th International Symposium on Remote Sensing of the Environment and the 18th Annual Symposium of the Canadian Remote Sensing Society* (Canadian Remote Sensing Society), pp. 401–404.

Ranchin, T., and Wald, L., 1996c, Preparation of SPOT 4 mission: Production of high resolution (10m) multispectral images using the ARSIS method, In *Proceedings of the 15th Symposium of EARSeL, Progress in Environmental Research and Applications*, edited by E. Parlow, EARSeL (Rotterdam: A.A. Balkema), pp. 175–179.

Ranchin, T., and Wald, L., 1997, Fusion d'images HRV de SPOT panchromatique et multibande à l'aide de la méthode ARSIS: apports à la cartographie urbaine, In *Actes des sixièmes journées du éseau de télédétection de l'UREF: Télédétection des milieux urbains et périurbains*, UREF (Montréal: AUPELF), pp. 283–290.

Ranchin, T., Wald, L., and Mangolini, M., 1994, Efficient data fusion using wavelet transforms: the case of SPOT satellite images, In *Mathematical Imaging: Wavelet Applications in Signal and Image Processing, Proceedings of SPIE International Symposium on Optics, Imaging and Instruments* (SPIE), pp. 171–178.

Ranchin, T., Wald, L., and Mangolini, M., 1996, The ARSIS method: a general solution for improving spatial resolution of images by the means of sensor fusion, In *Proceedings of the conference on Fusion of Earth data: merging point measurements, raster maps and remotely sensed images*, edited by T. Ranchin, and L. Wald, SEE/URISCA (Nice: SEE/URISCA), pp. 53–58.

Rioul, O., and Vetterli, M., 1991, Wavelets and signal processing. *IEEE Signal Processing Magazine*, **8**, 14–38.

Wald, L., and Ranchin, T., 1995, Fusion of images and raster-maps of different spatial resolutions by encrustation: smoothing the edges. *Computers Environment and Urban Systems*, **19**, 77–87.

CHAPTER 3

Urban Pattern Characterization through Geostatistical Analysis of Satellite Images

Pietro Alessandro Brivio and Eugenio Zilioli

3.1 Introduction

Metropolitan agglomerations and urban settlements are the most intensive forms of land use yet devised. In recent years, some of the most drastic alterations to local climate (microclimate) have been associated with them, with consequent impacts on the health and quality-of-life of their human inhabitants. At the beginning of this century urban areas represented only 20% of the total human population: in the hundred years since then this figure has increased to almost 50%. In fact, current estimates suggest that 42% of the world's population are now urban dwellers, with local variations ranging from as much as 72% on average in more-developed countries, to 34% in some less-developed countries (World Population Reference 1993).

In metropolitan and urban areas, the problems relating to the rapid transformations that are taking place in terms of land cover and land use are now very pronounced. As a result, the availability of detailed, timely information on urban areas is of considerable importance both to the management of current urban activities and to forward planning (Fabos and Petrasovits 1984). Satellite remote sensing has the potential to provide some of the necessary information and the body of literature concerned with urban applications of this technology continues to grow, with notable contributions on rural-to-urban land conversion (Jensen 1983, Forster 1985, Gomarasca *et al.* 1993), housing density and population estimation (Ilisaka and Hegedus 1982, Lo 1993), and urban climates (Henry *et al.* 1989, Kawashima 1994, Brivio *et al.* 1995). There is, however, a number of difficulties associated with the complexity of the urban landscape — due to the extreme heterogeneity of surface materials (land cover) at both the inter-pixel and intra-pixel scales — which demand the use of textural information in addition to usual multispectral data (Webster and Bracken 1993, see also Pesaresi and Bianchin this volume).

It is important at the outset to distinguish between texture and pattern. Here, we take image *texture* to refer to the frequency of tonal change, while *pattern* refers to the spatial arrangement of textural components. Although it is common to distinguish between them, texture and pattern share two important attributes, namely a spatial extension of some local 'order' and a repetition (which may be to a greater or lesser degree random) of elementary parts (Pratt 1978). Consequently, different aggregations of the same basic scene elements can combine to produce various different textures and patterns, each of which reflects the different underlying spatial organizations of the urban scene. Numerous quantitative measures of image texture have been developed, with the aim of capturing this underlying spatial organization. Many of these were originally developed and applied in the field of computer vision, but these have subsequently been augmented and extended for remote sensing applications. The most widely-used approaches to the extraction of textural and structural features from digital images include grey-level spatial dependency or 'co-occurrence' matrices (Haralick *et al.* 1973), grey-level difference vectors (Weszka *et al.* 1976), Fourier spectrum analysis (He *et al.* 1988), fractal geometry (Peleg *et al.* 1984) and geostatistical analysis (Woodcock *et al.* 1988a, Woodcock *et al.* 1988b). It is the last of these which is considered in detail in this Chapter and, in particular, the semi-variogram.

Initially developed as a tool for mining exploration (David 1977, Clark 1979), the semi-variogram has recently been adopted for use in the analysis of remotely-sensed images. Examples of its application include the analysis of data from ground-based radiometry (Curran 1988, Webster *et al.* 1989), the investigation of forest canopy structure (Cohen *et al.* 1990) and in scene modelling and digital image-processing (Ramstein and Raffy 1989, Brivio *et al.* 1993b). In this study, we apply the same geostatistical techniques (semi-variograms) to explore the spatial patterns expressed in Landsat Thematic Mapper (TM) images of urban areas. More specifically, we use examples from various parts of Italy which, visually at least, reflect the rather different spatial arrangement of urban settlements in the northern and southern parts of the country. Two approximations to the theoretical bi-dimensional semi-variogram are explored, namely the *multi-directional* and the *matrix* semi-variograms. The former provides a convenient way of describing the spatial structure of the image (and, hence, the corresponding scene) in terms of its prevailing directional patterns or anisotropy. The latter allows a simple parameterization of texture of different spatial samples drawn from one or more images (scenes). Finally, a comparative analysis is conducted to examine the influence of scale changes (i.e. sensor spatial resolution) on spatial pattern characterized by geostatistical techniques.

3.2   The Structure Function $\gamma(h)$

A digital image of the Earth surface derived by means of a remote-sensing device can be thought of as a set of radiometric measurements, each associated with a particular geographical location. In mathematical terms, image brightness $B(x)$ is a function of the two-dimensional spatial location x of the measurement and can be considered to be the realization of a 'regionalized variable' (Matheron 1965). A key concept to take forward at this stage is the basic assumption of spatial dependence in the data; that is, observations located close together in terms of their geographical coordinates are expected to be more similar to each other than observations located further apart. This spatially dependent variation, inherent in all digital images, can be analyzed by using the properties of the semi-variogram, also called structure function $\gamma(h)$.

The semi-variogram $\gamma(h)$, which is a function relating the semi-variance to the directional distance between two samples, can be expressed through the relationship:

$$\gamma(h) = \frac{1}{2(n-h)} \sum_{i=1}^{n-h} [B(x_i) - B(x_i + h)]^2 \tag{3.1}$$

where

$B(x_i)$ is the radiance value measured at pixel (x_i),

h is the lag, or distance in the image, expressed as a number of pixels and defining the different locations $(x_i + h)$ at which the regionalized variable B is observed, and

n is the number of observations used to estimate $\gamma(h)$.

For any given data set, the shape of the semi-variogram can be determined empirically using Equation 3.1. The form of such *experimental* semi-variograms can then be compared to a number of theoretical models. More precisely, experimental semi-variograms are usually fitted to a set of bounded and unbounded models (Clark 1979, Webster 1985, Davis 1986), among which the ones that are generally of greatest interest are the exponential model (Equation 3.2), the spherical model (Equation 3.3), the power model (Equation 3.4) and de Wij's model (Equation 3.5):

$$\gamma(h) \;=\; c\left[1 - exp^{-\frac{h}{a}}\right] \tag{3.2}$$

$$\gamma(h) \;=\; c\left[\frac{3}{2}\left(\frac{h}{a}\right) - \frac{1}{2}\left(\frac{h}{a}\right)^3\right] \quad \text{for } h < a \tag{3.3}$$

$$\gamma(h) \;=\; c \quad \text{for } h > a$$

$$\gamma(h) \;=\; w\,h^a \quad \text{for } 0 < a < 1 \tag{3.4}$$

$$\gamma(h) \;=\; 3a\log h \tag{3.5}$$

A number of important parameters can be derived by fitting the experimental semi-variogram to one of these theoretical semi-variograms, notably the nugget, sill, range, derivative at the origin, and degree of anisotropy (Curran 1988, Jupp *et al.* 1989). The *sill, c,* indicates the value of the semi-variogram at which the function stabilizes to form a plateau as separation (*lag*) distances increase; it is equal to the general data variance. The *range, a*, corresponds to the lag, *h*, at which the semi-variogram reaches the sill, *c*; this is the critical distance at which the correlation structure ceases to exist and beyond which the data vary randomly. As the lag, *h*, tends to zero, the semi-variogram function does not always intercept the origin. This indicates that there is a component of spatially-independent variance in the data, perhaps due to the noise of the sensor system itself or to random surface variation. This is known as the *nugget* effect.

3.3   Geostatistical Analysis of Satellite Sensor Images

Satellite sensor images are two-dimensional representations of the observed spectral radiance emanating from the Earth surface. The correct formulation of the semi-variogram is therefore a two-dimensional function $\gamma(p, q)$, where p and q are the lags in each of the two dimensions, respectively. In other words, the semi-variogram function depends not only on the distance, but also the direction, θ, in which the semi-variance is estimated. Two-dimensional (2D) semi-variograms are commonly used to reveal anisotropy in the observed surface which may, for example, be related to the orientation of objects in the scene or to the direction of the solar illumination. They are, however, more difficult to interpret in terms of their shape, range of influence and height relative to the estimate of the sill (Woodcock *et al.* 1988b). Nevertheless, the form of 2D semi-variograms is usually a geometric extension of the corresponding 1D forms, such that a cylindrical projection of the sample values, resulting a sheaf of lines, will often serve quite well. The 2D function $\gamma(p, q)$ can also be approximated reasonably well by the *multi-directional* semi-variogram (Webster 1985).

When ground features do not show any evident directional organization, the structure of the scene can be said to be isotropic. In this case, the $\gamma(h)$ function can be calculated by the matrix method. This method initially evaluates the semi-variogram separately in the row-direction and column-direction of the image and then takes the average of these two semi-variograms at each lag, *h* (Cohen *et al.* 1990). Although this technique does not account for anisotropic variation in the images, the shape of matrix semi-variogram — derived by collapsing 2D information into a single semi-variogram — maintains the anisotropic behaviour of the full 2D semi-variogram (Woodcock *et al.* 1988b). Moreover, this technique is an effective tool for quantitative analysis and parameterization purposes; consequently, we have adopted this approach for the present study, which examines a number of Landsat-TM images, each having a ground resolution element (pixel size) of $30m \times 30m$ in the six optical (visible to middle infrared) spectral wavebands.

3.3.1   Examining the Spatial Pattern of Contrasting Terrestrial Landscapes

We start by analyzing the spatial organization of different terrestrial landscapes using three large sample areas, each about $80km^2$ (or 300×300 pixels) in size. These areas were selected to represent different levels of human influence on the the spatial structure of the landscape (Table 3.1):

Table 3.1: General description of the main features found in the three test areas used in this study.

Scene Sample	Site	Feature Description	Spectral Band
Mountainous	Alps	Natural landscape	TM 4 ($0.76\mu m$–$0.9\mu m$)
Rural	Palmanova	Scattered settlements	TM 4 ($0.76\mu m$–$0.9\mu m$)
Urban	Milan	Urban environment	TM 3 ($0.63\mu m$–$0.69\mu m$)

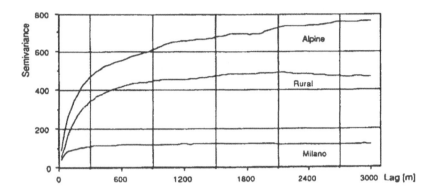

Figure 3.1: Experimental semi-variograms at landscape level for three different sites in North Italy from Landsat-TM imagery.

1. The metropolitan agglomeration of Milano. This area has one of the highest population densities in Italy, as well as a high concentration of industrial activities and a strongly developed communications infrastructure (i.e. railway, motorway and road network);

2. The rural area surrounding Palmanova (Friuli) in north-east Italy. This area comprises numerous small settlements scattered throughout a traditionally agricultural region. Recently, the area has experienced a rapid and continuing process of urbanization;

3. A natural/semi-natural mountainous environment in the Alps. In this area, the spatial structure reflects the terrain morphology, and is marked by several different drainage networks, ranging in form from dendritic, through sub-parallel, to irregular.

Different spectral wavebands were selected for use with each of the three study areas, based on previous experience. The near-infrared channel (TM band 4) has been found to be the most effective when dealing with natural/semi-natural and agricultural landscapes, and as such was selected for use in the context of the mountain range and rural/agricultural sites. In our experience, the red waveband (TM band 3) is better adapted to distinguish different artificial features (Brivio *et al.* 1993a, Brivio *et al.* 1994), and consequently this channel has been used to define the urban agglomeration.

The results of the geostatistical analysis of the three study areas are reported in Figure 3.1. Visual inspection of the experimental semi-variograms shows two broad shapes: one for

Table 3.2: Summary description of the three sample areas, each about $7km^2$ or 90×90 pixels in size, selected from the Landsat TM image covering the Po Plain, northern Italy. Note that the last column describes the percentage of the population in the corresponding province living in urban areas.

Town	Spatial Structure Description	Population	% of Province
Milano	Radial and concentric patterns	1 600 000	39.5
Torino	Regular square pattern	1 100 000	47.6
Novara	Small city, no evident pattern	100 000	20.1

which the semi-variograms reach a clear sill (the rural and urban sites), the other where no sill is reached (the natural alpine site). In the former case, where the landscape is influenced by the presence and activities of human beings, the semi-variance reach the sill at a lag distance of 150–300m and 900m for the urban and rural sites, respectively. However, in the mountainous area which is largely unaffected by anthropogenic activity, the semi-variogram appears to be unbounded and the semi-variance continues to increase with ever larger lag distances.

3.3.2 Models of Urban Structure

Italy is sometimes called 'the country of a thousand bell towers'. In fact, the Italian landscape — the product of many different cultures over the years — is strewn with numerous different types of urban areas, ranging from small villages and ancient towns to large metropolitan areas. The historical processes, notably those dating from the Romans and Middle Ages through to the Renaissance, have generated contrasting models of spatial organization and urban structure. Each town is therefore characterized by its own particular configuration, which is a summary of its history, as well as the influence of more recent developments and a range of environmental factors, such as the presence of water, its topographic relief, and so on.

In order to define some models of urban structure, three further sample areas were selected from within the Po Plain, in northern Italy. These correspond to the urban agglomerations of Milano, Torino and Novara, which are taken to indicate, respectively, linear, rectangular and circular/radial patterns of urban spatial organization (Figure 3.2). A summary description of the three sub-scenes, each about $7km^2$ or 90×90 pixels in size, is given in Table 3.2, including the urban population and the percentage that this represents with respect to the total population in that province. All three sub-scenes were selected from the same Landsat TM image (scene number 193-28), which was acquired on June 18, 1991, such that the effects of differences in atmospheric conditions and other confounding factors between the sub-scenes is effectively minimized.

As expected, the results of the geostatistical analysis of these three sub-scenes, shown in Figure 3.3, suggest that the height of the sill in the semi-variograms is related to the spatial organization of the corresponding urban scenes. Torino, for example, which has preserved the regular grid of streets typical of the Roman era, has the smallest maximum value of semi-variance due to its more organized spatial structure. Milano, on the other

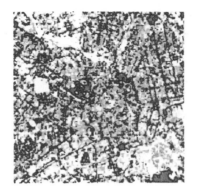

(a) Milano (image)

(b) Torino (image)

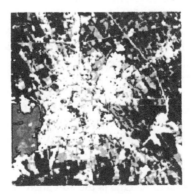

(c) Novara (image)

Figure 3.2: Satellite images of three urban sample areas.

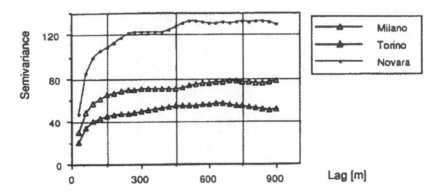

Figure 3.3: Experimental semi-variograms for the three urban areas exhibiting different types of spatial organization.

hand, which developed without any such morphological constraints, displays higher semi-variance values. The semi-variance values for Novara are the highest of all, reflecting the complex spatial structure of this urban area.

Exploring Anisotropy in Urban Patterns

It should be evident from the satellite sensor images in Figure 3.2 that most natural and anthropogenic environments exhibit spatial patterns that are far from isotropic. In such circumstances, the semi-variogram function depends not only on the distance, but also on the direction, θ, in which the semi-variance is estimated. In this study we have employed the multi-directional semi-variogram to explore the degree to which our three urban sub-scenes display such anisotropy. The result of applying this technique to each of the corresponding images is a sheaf of experimental data whose envelopes may be grouped, to a greater or lesser degree, in relation to the anisotropic variation revealed along the directions explored. The following eight directions, θ, were used:

W-E	(0°)	WNW-ESE	(−26.56°)
NW-SE	(−45°)	NNE-SSW	(−116.56°)
N-S	(90°)	NNW-SSE	(−63.44°)
NE-SW	(−135°)	ENE-WSW	(−153.44°)

Several parameters have been developed that can be used to analyze the degree of anisotropy in scene spatial structure. One is the so-called A/B anisotropy ratio; where A and B are the gradients of the semi-variograms in the direction of maximum and minimum variation, respectively (Webster and Oliver 1990). A second measure, suggested by Brivio *et al.* (1994), is the angle Δ between the maximum (φ) and minimum (ψ) gradients of the experimental, multi-directional semi-variograms. A third is the coefficient $\Omega(\theta)$, which is

Table 3.3: Anisotropy parameters for the three urban sub-scenes (image window = 90×90 pixels). A = maximum gradient; B = minimum gradient; A/B = ratio of anisotropy; φ = maximum angle; ψ = minimum angle; $\Delta = \varphi - \psi$; θ = angle of max gradient; χ = angle of minimum gradient; $\Omega(\theta)$ = coefficient of anisotropy in direction θ.

Site	Milano	Torino	Novara
A	1.2126	0.8852	2.3704
φ	50.48	41.51	67.12
θ	N-S	W-E	W-E
B	0.5959	0.3935	0.7240
ψ	30.79	21.47	35.90
χ	NNE-SSW	NNW-SSE	ENE-WSW
A/B	2.03	2.24	3.27
Δ	19.69	20.03	31.22
$\Omega(\theta)$	1.00	0.71	1.13

derived using the following equation (David 1977):

$$\Omega(\theta) = \sqrt{A^2 cos^2(\theta - \varphi) + B^2 sin^2(\theta - \varphi)} \tag{3.6}$$

This parameter indicates the degree of affine anisotropy in the image in a specific direction, θ. Values of these anisotropy parameters have been calculated for each of the three urban sub-scenes and the results are presented in Table 3.3.

Scaling Effects

The importance of the spatial scale of observation has been identified by Marceau *et al.* (1990), in a study which explored the discriminating power of various texture measures applied to images acquired at different levels of sensor spatial resolution. In the extreme case, one can imagine an urban area covered by a single pixel, such that we are unable to capture any information about the pattern of building lots, local road networks, and so on.

To explore the effect of spatial scale on the geostatistical techniques that have been introduced thus far, we have performed a series of experiments on a sub-scene extracted from a Landsat TM image of Milano. The image was re-sampled in four successive steps, increasing the pixel size (i.e. reducing the effective sensor spatial resolution) at each step. Thus, the size of the image was reduced from the original 300×300 pixels at 30m resolution, to 150×150 pixels at 60m resolution, and so on, with the final image being 18×18 pixels in size and having an effective spatial resolution of 480m. The resampling scheme simply involved taking the mean value of the four pixels included in the resolution cell at the next (coarser) spatial scale. Experimental semi-variograms were then calculated for these data (Figure 3.4). The results suggest that the experimental semi-variograms maintain a similar shape (note, in particular, the local minimum value at large lag distances; $h \approx 2200m$) across the range of spatial scales considered here (30m–500m), although there is a progressive reduction in the height of the sill, due to the smoothing effect of re-sampling procedure.

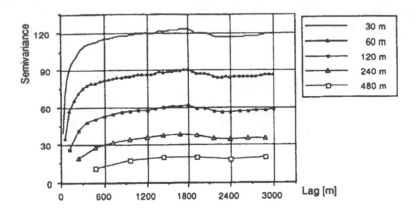

Figure 3.4: Semi-variograms of the Milano urban area as a function of sensor spatial resolution (30m–500m).

Table 3.4: General description of the five small small sub-scenes used to examine the local-scale image texture indicative of different types of urban area. The symbols refer to the codes used in Figure 3.5.

Site	Sample Feature Description	Symbol
Siracusa	Old town centre	URB/Sir
Siracusa	Suburban area	SUB/Sir
Floridia	Small town	URB/Flo
Priolo	Large industrial district	INL/lar
Priolo	Small industrial district	INL/sma

Urban Texture at Local Scale

The geostatistical approach was also applied at the local scale to Landsat TM image sub-scenes covering a range of industrial areas and urban centres (Table 3.4). Five windows of 24×24 pixels (i.e. $720m \times 720m$) were extracted from an image covering Sicily, southern Italy, namely: two industrial districts close to Priolo, one with predominantly large building units, the other with smaller building units, but each indicative of the recent and sometimes uncontrolled industrialization in this area; two residential areas, one at the fringe and one in the centre of the coastal town of Siracusa; and the small, compact town of Floridia, $15km$ away.

Data from Landsat TM band 1 ($0.45\mu m$–$0.52\mu m$, blue-green) were selected for this exercise since they provide good spectral separation between the man-made features of interest. The experimental semi-variograms for each of the sample areas derived from these data are presented in Figure 3.5. It is evident from this figure that the urban centres show 'flattened' semi-variograms, with maximum variances typically much lower than, for exam-

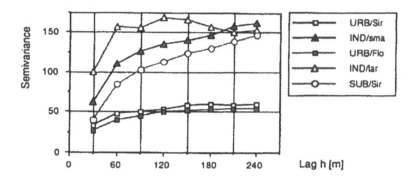

Figure 3.5: Plots of the sample semi-variograms obtained for each of the five sub-scenes extracted from a Landsat TM image (Channel 1) centred on Siracusa, southern Italy. See Table 3.4 for an explanation of the key.

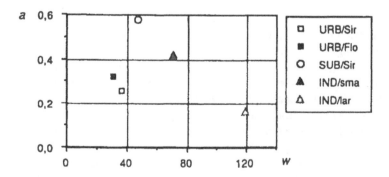

Figure 3.6: Plot of $(w{-}a)$ parameter space for five urban sub-scenes. The values of w and a were obtained by fitting experimental semi-variograms derived from these sub-scenes to a power model.

ple, the two industrial sites. The residential area at the urban fringe of Siracusa (SUB/Sir) occupies an intermediate position between these two extremes, although it is closer in form and magnitude to the industrial site with smaller building sizes (IND/sma). This is due to the higher variances created by the presence of the spectrally-contrasting intra-urban vegetation found at the urban fringe. Finally, the highest values of semi-variance occurs in the large-building industrial site (IND/lar), which appears to show a degree of periodicity beyond $60m$–$90m$. This distance corresponds to the approximate size of the petroleum refinery plants found in this sub-scene.

In order to standardize the results described above, and to permit a somewhat more quantitative analysis, the experimental semi-variograms were fitted to the theoretical models described in section 3.2. Based on previous experience (Brivio *et al.* 1993a), the power (or generalized linear) model was adopted in this study, because it offers a convenient compromise between simplicity and flexibility in terms of its ability to describe different semi-variogram shapes. Experimental semi-variograms were fitted to the power model using the least-squares method. This provided estimates of the two parameters, w and a. The goodness-of-fit between the experimental and theoretical semi-variograms was evaluated using the correlation coefficient, r, which was found to exceed 0.95 for all sites, with the exception of the large-building industrial district for which $r = 0.7$. The parameter values were then plotted in $(w$–$a)$ space (Figure 3.6). This suggests that $(w$–$a)$ space can be used to distinguish different types of urban area. For example, the urban centres of Siracusa and Floridia are located close together in this plot and are largely separate from the other three sites examined. Likewise, the industrial district with large buildings occupies a distinct and separate position within $(w$–$a)$ space in the bottom-right corner of the plot. The suburban area and the small-building industrial district also appear to occupy discrete regions of $(w$–$a)$ space.

3.4   Conclusions

Considerable effort has been expended by digital photogrammetrists on the identification of individual buildings in images of urban areas, and by remote sensing specialists on mapping and monitoring urban land cover/land use from satellite sensor images (see Mesev *et al.*, Chapter 5 and Barnsley *et al.*, Chapter 7 this volume). Much less work has, however, been directed towards the description and quantification of the structure and texture of urban areas in digital remotely-sensed images. Due to space constraints, the methods and results presented in this Chapter can only give a general flavour of the possible contribution that an analysis of texture can bring to better understand urban form and, hence, the relationships between pattern and process. However, we have demonstrated the application of geostatistical analysis to various urban sub-scenes extracted from Landsat TM images. These were selected as examples of typical Italian urban areas, ranging from large metropolitan agglomerations to small towns. The results of this analysis provided an insight into how these techniques might be used to distinguish and interpret the spatial pattern of different types of urban structure. It was suggested that remotely-sensed images of urban areas often exhibit a degree of anisotropy in terms of their observed spatial variation, due to the presence of directional artefacts (e.g. transportation networks) in the corresponding urban scene. Multi-directional, experimental semi-variograms were used to explore and quantify this anisotropy and, hence, to differentiate between urban areas exhibiting different spatial structures. The effects of the spatial scale of observation on the ability to detect characteristic urban patterns were also analyzed by simulating image acquisition at different sensor spatial resolutions. Finally, focusing in on the local scale, experimental semi-variograms were fitted to a power model in order to extract quantitative measurements of image texture over different types of urban area and, hence, to distinguish between them.

3.5   References

Brivio, P. A., Doria, I., and Zilioli, E., 1993a, Application of semivariograms to satellite digital images for territorial analysis, In *Image Processing: Theory and Applications*, edited by G. Vernazza, A. N. Venetsanopoulos, and C. Braccini (Amsterdam: Elsevier), pp. 385–388.

Brivio, P. A., Doria, I., and Zilioli, E., 1993b, Aspects of spatial autocorrelation of Landsat-TM data for the inventory of waste disposal sites in rural environment. *Photogrammetric Engineering and Remote Sensing*, **59**, 1377–1382.

Brivio, P. A., Doria, I., and Zilioli, E., 1994, Structure function analysis of natural scenes from Landsat TM data. *ITC Journal*, **1**, 1–6.

Brivio, P. A., Genovese, G., Massari, S., Mileo, N., Saura, G., and Zilioli, E., 1995, Atmospheric pollution and satellite remotely sensed surface temperature in metropolitan areas, In *Proceedings of the EARSeL Workshop on Pollution Monitoring and GIS*, EARSeL, pp. 40–46.

Clark, I., 1979, *Practical Geostatistics* (London: Applied Science Publishers).

Cohen, W. B., Spies, T. A., and Bradshaw, G. A., 1990, Semivariograms of digital imagery for analysis of canopy structure. *Remote Sensing of Environment*, **43**, 167–178.

Curran, P. J., 1988, The semi-variogram in remote sensing: an introduction. *Remote Sensing of Environment*, **24**, 493–507.

David, M., 1977, *Geostatistical ore reserve estimation*, Developments in Geomathematics, 2 (Oxford: Elsevier).

Davis, J. C., 1986, *Statistics and data analysis in geology* (New York: John Wiley and Sons).

Fabos, J. G., and Petrasovits, I., 1984, Computer-aided land use planning and management, Research Bulletin 693, Massachusetts Agricultural Experiment Station, 25–45.

Forster, B. C., 1985, Principal and rotated component analysis of urban surface reflectances. *Photogrammetric Engineering and Remote Sensing*, **51**, 475–477.

Gomarasca, M. A., Brivio, P. A., Pagnoni, F., and Galli, A., 1993, One century of land-use changes in the metropolitan-area of Milan (Italy). *International Journal of Remote Sensing*, **14**, 211–223.

Haralick, R. M., Shanmugan, K., and Dinstein, I., 1973, Textural features for image classification. *IEEE Transactions on Systems, Man and Cybernetics*, **8**, 610–621.

He, D.-C., Wang, L., and Guilbert, J., 1988, Texture discrimination based on an optimal utilization of texture features. *Pattern Recognition*, **2**, 141–146.

Henry, J. A., Dicks, S. E., Wetterqvist, O. F., and Roguski, S. J., 1989, Comparison of satellite, ground-based, and modeling techniques for analyizing the urban heat island. *Photogrammetric Engineering and Remote Sensing*, **55**, 69–76.

Ilisaka, J., and Hegedus, E., 1982, Population estimation from Landsat imagery. *Remote Sensing of Environment*, **12**, 259–272.

Jensen, J. R., 1983, Urban/suburban land use analysis, In *Manual of Remote Sensing, Second Edition*, edited by R. N. Colwell (Falls Church, Virginia: American Society of Photogrammetry), pp. 1571–1666.

Jupp, D. L. B., Strahler, A. H., and Woodcock, C. E., 1989, Autocorrelation and regularization in digital images. II: Simple image models. *IEEE Transactions on Geoscience and Remote Sensing*, **27**, 247–258.

Kawashima, S., 1994, Relation between vegetation, surface temperature and surface composition in the tokyo region during winter. *Remote Sensing of Environment*, **50**, 52–60.

Lo, C. P., 1993, A GIS approach to population estimation in a complex urban environment using SPOT multispectral images. *International Archives of Photogrammetry and Remote Sensing*, **29**, 935–941.

Marceau, D. J., Howarth, P. J., Dubois, J. M. M., and Gratton, D. J., 1990, Evaluation of the gray-level co-occurrence matrix-method for land- cover classification using SPOT imagery. *IEEE Transactions on Geoscience and Remote Sensing*, **28**, 513–519.

Matheron, G., 1965, *Les variables régionalisées et leur estimation* (Paris: Masson).

Peleg, S., Naor, J., Hartley, R., and Avnir, D., 1984, Multiple resolution texture analysis and classification. *IEEE Transactions on Pattern Analysis and Machine Intelligence*, **6**, 518–523.

Pratt, W. K., 1978, *Digital Image Processing* (New York: John Wiley and Sons).

Ramstein, G., and Raffy, M., 1989, Analysis of the structure of radiometric remotely-sensed images. *International Journal of Remote Sensing*, **10**, 1049–1073.

Webster, C. J., and Bracken, I. J., 1993, Exploring the discriminating power of texture in urban image analysis. *International Archives of Photogrammetry and Remote Sensing*, **29**, 942–948.

Webster, R., 1985, *Quantitative analyis of soil in the field* (New York: Springer-Verlag).

Webster, R., Curran, P. J., and Munden, J. W., 1989, Spatial correlation in reflected radiation from the ground and its implications for sampling and mapping by ground-based radiometry. *Remote Sensing of Environment*, **29**, 67–78.

Webster, R., and Oliver, M. A., 1990, *Statistical Methods in Soil and Land Resource Survey*, Spatial Information Systems (Oxford: Oxford University Press).

Weszka, J. S., Dyer, C. R., and Rosenfeld, A., 1976, A comparative study of texture measures for terrain classification. *IEEE Transactions on Systems, Man and Cybernetics*, **6**, 269–285.

Woodcock, C. E., Strahler, A. H., and Jupp, D. L. B., 1988a, The use of variograms in remote sensing: I. Scene models and simulated images. *Remote Sensing of Environment*, **25**, 323–343.

Woodcock, C. E., Strahler, A. H., and Jupp, D. L. B., 1988b, The use of variograms in remote sensing: II. Real digital image. *Remote Sensing of Environment*, **25**, 349–379.

World Population Reference, 1993, *World population data set* (Washington, D.C.: Population Reference Bureau).

CHAPTER 4

Recognizing Settlement Structure using Mathematical Morphology and Image Texture

Martino Pesaresi and Alberta Bianchin

4.1 Introduction

Mathematical morphology provides an explanatory tool and a guide for recognizing the complex phenomena which generate observed spatial structures. In the first part of this Chapter the origins of the morphology concept are investigated in relation to evolutionary theory and the thoughts of the Italian Architectural School. Image processing techniques are also viewed as an aid to the construction of an analytical 'urbanistic' discipline. The morphological/textural approach to remotely-sensed image processing is then examined as potential means of extracting spectral and spatial descriptors of an image. Two concepts of urban space are introduced: (i) marked, or labyrinthine, space and (ii) homogeneous space. In the second part of the Chapter, the role of structural information derived from satellite-sensor images as a strategic resource in the analysis of human settlement is examined from two different perspectives:

1. *Operational*: concerning the ways in which structural information can be used to improve the accuracy of measures of the shape and extent of built-up areas; and,

2. *Methodological*: concerning the isomorphism between morphological/textural image processing techniques and the language of architectural design.

4.2 Origins of the Morphology Concept

The study of 'urbanism' is concerned with the relationships between human-kind (or society) and physical space. Today's planning and urban theory has its origins in the social conflicts generated by the Industrial Revolution, and subsequent attempts by the state to resolve them. Study of the social patterning of urban areas developed in the Nineteenth Century around typologies of the structure and configuration of society. Early analysis of urban morphology developed around biological analogies, and the way in which different types of organisms are arranged into classes and sub-classes. As such, the study of urban morphology is rooted in Evolutionary Theory, and borrows from the zoological and biological sciences which sought to develop phylogenetic classifications to account for the interdependencies between different animal species. Morphological analysis was intended to be an explanatory tool, a means of relating spatial form to generating process (see also Longley and Mesev, Chapter 9 this volume).

Morphological analysis involves the aggregation of elements, none of which may be important in isolation. In order to develop a classification, the significance of elements arises from their association with other elements, which may be more or less important in the scheme of things. Darwin underscored a number of points which should be borne in mind when we consider the morphological approach:

- genealogy holds the key to identification and hence to an understanding of morphology;

- analysis of morphology through analogy often provides weaker explanatory power than a study of genealogy; and

- morphological analysis assumes equifinality of outcome — that is, similar shapes share common historical origins and processes.

Taken together, these concepts provide the explanatory power of morphological analysis.

Form and structure are elements of morphology, since both relate to the organization of parts within an overall scheme. The term 'social morphology' nevertheless has many possible meanings. For some, it pertains to the 'physical appearance of social reality' — something visible which pertains to invisible phenomena. From among these visible structures, urbanists are interested in spatial distributions and in the ways in which society takes shape across space. Such distributions, structures, shapes and schemes are taken by analytical planners to provide visual clues to the evolution of space. Architects of the School of Aymonino and Rossi, on the other hand, use these formal shapes to frame project activity.

What is crucial is how we might identify the 'real' structures, those constant elements of organization that offer systematic explanatory power. Darwin suggested that genetics offered one of the more reliable means of ascertaining significant meanings, as has Conzen (Whitehand 1981) in the field of urban geography. Yet such an approach is inevitably restricted to retrospective analysis, and is useless for prospective planning purposes. Within the urban realm, the definition of structure is also somewhat subjective and, although guided by some general rules, is largely intuitive. We can only judge which definition is the most useful in terms of its explanatory power, or in terms of its internal consistency (Ginzburg 1981). Furthermore, such reliance upon essentially intuitive methodology, rather than clearly-defined technical tools, makes it impossible to compare objectively successive states of an urban system over time.

4.2.1 From 'Urban Form' to Morphology

In linking the static description of urban form to the systematic analysis of urban morphology, we will adopt the perspective of the urbanist-architects. That is to say, we will consider the city as the locus of urban phenomena, not only in order to understand it, but also to draw out 'some theoretical clues which, once defined and verified, can be assumed as data useful to check the consistency of the urban project in the contemporary city' (Aymonino 1977, p.53).

This is the position expressed by a group of 1960s Italian architects, which acknowledged that the city no longer developed according to predefined city-wide rules (Aymonino 1977), and that the urban whole was more than the sum of its architectural parts. While in the past the quality of a single monument might have been the fulcrum of the organization of an entire city (or of parts of it), in the contemporary city this notion of quality has been superseded. Today, quality is defined by the quantity of buildings, the character of which is manifest in the way they are structured according to a set of simplified, but not predefined, rules. Today's city is not recognizable in its existence through meaningful architectural edifices. Private land ownership and speculative activity has led to the primacy of size over form. The city, no longer an urban form, has become a massive agglomeration (Aymonino 1977, p.51). Nevertheless, Aymonino (1977) develops three important points:

1. the central object of theoretical thinking should be the physical and spatial form of the city;

2. there should be an analytical, and if possible scientific, basis to the analysis of urban form; and,

3. the goal of urban morphology research is to unify the study of architecture and urbanism in one discipline, where architecture can develop its analytical character and urbanism its interest in physical and spatial structure.

4.2.2 The Contribution of Image Processing to the Development of an Analytical Discipline

The three points developed by Aymonino, outlined above, suggest some possible rôles for remote sensing in the field of urban morphology analysis, as well as some scientific grounds on which to develop analytical tools and conceptual categories.

In order to illustrate this, we might draw a comparison with the rôle of statistics in the development of the study of demography. Today it is not possible to imagine a study of demography without the support given to it by statistics. This support is through the methodology which allows the calculation of indicators, parameters and descriptive statistics which, in turn, shape conceptual categories. Concepts such as population structure are closely related to the methods used to derive them.

Yet the concept of population structure is not unambiguous, since it implies a system of relationships which governs principles. It is, rather, a concept which varies in its totality, and the shape it takes depends on the way in which entities and relationships are defined.

Demographic entities are not necessarily related to physical objects, but rather are abstractions arising in part out of the way in which they are investigated. Entities must, however, satisfy some general requirements:

- they must be consistent with the phenomenon under investigation;

- they must aid, not hinder, explanation of the phenomenon under investigation; and,

- they must facilitate comparison between entities.

The creation of structure always requires abstraction, and such abstraction should provide a set of guiding principles which select, organize and order relevant elements, independent of contingent factors. Yet this assertion risks tautology: it is always necessary to specify a method of (often discipline-specific) analysis at the same time as a means of defining the relevant variables.

The lack of rigour in definition and analysis also arises in part because of the paucity of spatial analysis in providing measures of the distribution of physical and other spatial structures on the ground. As we will see below, remote sensing provides a number of new measures of urban structure.

4.2.3 Two Kinds of Spatial Information in the Mind of the Town Planner

In 1964, the famous Italian architect-urbanist G.C. De Carlo, formerly the scientist responsible for PIM[1], prepared the final report on the methodology to be used in the inter-municipal plan (De Carlo 1966). This application is well known in Italy as an exemplar of the principles of Systems Theory applied to town and regional planning. It was characterized by prominent use of the latest methods and techniques — such as optimization and (punch card) computer-based analysis. There is some resonance between this study and today's use of satellite data to study urban areas, not just in terms of problem formulation, but also the likely outcomes of analysis. More specifically, these concern the representation of space and the nature of spatial information processing.

The PIM model was based upon a grid-based data structure[2]. The remit of the project went beyond mere representation, however. The PIM staff set out to use automated techniques to create spatial information, using rules of grid connectivity to measure the compactness, porosity and dispersion of built areas. Thus 'compact' areas were defined as groups of ten or more contiguous cells, more than half of which had adjoining cells on two or more sides (De Carlo 1966, p.83).

This example illustrates that urbanists have long recognized the importance of size and topology. Moreover, it is also interesting to observe that urbanists already had the conceptual tools to describe satellite data, almost ten years before the launch of Landsat-1 (ERTS-1), even though they still do not use data from this source. It is not the intention here to elaborate on how this state-of-affairs has arisen — though it most probably reflects the differences in approach between architects and those working in economic and social disciplines.

---

[1] PIM is the Piano Intercomunale Milanese, the inter-municipal Plan of Milano, Italy.

[2] More than thirty eight thousand cards were punched, to represent a grid (the analogue of the present-day pixel) of resolution of $200m \times 200m$.

4.2.4 The Digital Image as a Formalized Reference System

The digital image can be thought as a formalized system of reference, as a means of ascribing place and position. In order to establish a general system of reference, we may posit the existence of the space E (a sub-set of all possible positions), of an object O (to which position may be ascribed) and of a subject S (an agent). It is then possible to characterize three empirical situations that are each radically different:

1. The case in which E is a marked space. This space is made heterogeneous by points and by zones that may be distinguished using a pre-defined classification. Place and position are identified by a search for reference points. It is necessary that the subject S has a relatively global vision of the space E.

2. The case where the space E is labyrinthine and the subject S does not have global vision. The subject can navigate only locally, using immediate reference points.

3. The case where the subject S has a global vision of the space E, yet the space E is 'not marked', does not possess points of reference, and is homogeneous.

In classical quantitative geography, 'the most instructive approach to spatial languages is to regard them as providing sets of rules governing the use of co-ordinate systems' . Moreover, 'the language of the spatial form' is 'a language of superposed coordinates to a homogeneous space'. This is the spatial-temporal language defined by Carnap at the end of the 1950s, in which objects are identified by the value of the measures of a list of associated properties. The hegemonic concept of space in the last thirty years in our discipline has been that of homogeneous space, as in the notion of distance and cost in the spatial economy. For some applications, however, homogeneous space is not the best choice: the recognition and description of urban areas using satellite data is perhaps one such example, as are the more general questions of spatial elaboration posed by the urbanist. A notion of labyrinthine space or of marked space is arguably more useful.

4.3 The Detection of Urban Areas using Satellite Data

As is well known, in terms of semantic structure, a digital image is much like an analogue photograph. The general problem is how to partition the digital image, and how to identify and segment its thematic categories.

Each digital image can be considered to be structured according to two kinds of primary descriptors — spectral and spatial. Traditionally, the image *classification* procedures used in remote sensing commonly work in the spectral domain and therefore make use of multispectral data, while image *segmentation* procedures involve spatial descriptors and typically make use of monospectral data.

4.3.1 Image Classification

If our problem is the automatic recognition of the built areas in a satellite image, the standard procedure is still multispectral classification, even though it is widely criticized when it is applied to high spatial resolution sensor data and where there exists high level of spectral variability within the candidate classes. High within-class variability is characteristic of urban residential areas, particularly in urban-rural transition areas (see also Weber, Chapter 8 this volume).

Image segmentation techniques have been used much less frequently to process satellite sensor data of urban areas. These techniques detect the boundaries (or contours) of objects through local transformations — such as gradient transformation — or through criteria based on measures of local homogeneity — as in the case of region growing (Zucker 1976) and split-and-merge algorithms (Haralick and Saphiro 1985). If the geometric resolution of the sensor approaches the size of the objects on the ground, then the objects constitute the (urban) thematic classes. This introduces texture effects into the image classification process, and may result in small, local discontinuities (around small isolated structures, or the thin structures alongside roads in semi-rural areas). Such effects cannot be considered as noise, yet they are handled only with difficulty by image segmentation methods.

Despite these issues, several authors have demonstrated that structural and radiometric information can lead to significant improvements in the classification accuracy of built-up areas. Various approaches have been tested, including texture indices (Flouzat *et al.* 1984, Marceau *et al.* 1990), local convolution transformations (Gong and Howarth 1990), fast morphological transformations and multi-sensor merging. There are trade-offs to be made, *inter alia* between the method of structural information handling, the nature of the raw data, the characteristics of the landscape of the built structures, and the robustness and processing time of the method, but it is important to understand that structural information is fundamental to automatic recognition of built-up areas.

Figure 4.1 shows the output of an automatic classification of Landsat TM and SPOT-HRV Panchromatic image data for the 'dispersive city' of the Venice region. It is derived from the DAEST-MURST research project "La città diffusa", developed during the 1990s within the Remote Sensing Unit of IUAV CICaF. The project study area measures $45km \times 45km$, includes the towns of Treviso, Castelfranco, Padova and Mestre-Venezia, and had a total population of *c.*1.5 million inhabitants in 1991. This region has a polycentric and diffused settlement structure, and as such provides a very stringent test for traditional image classification methods. Small and large urban nuclei, small accumulations along roads, and isolated buildings must all be carefully mapped in order to retain the fundamental information on settlement structure.

In this example, the estimated accuracy of classification built-up areas was more than the 90% at $30m$ spatial resolution. This result was obtained by using both spectral and structural information in a selective resampling procedure which merges the structural information of the SPOT-HRV Panchromatic data and with the spectral/radiometric classification of the Landsat TM data. The structural information is extracted using a transformation of the morphological gradient. This method is very fast and works well in identifying the urban texture of the Bassano and Cittadella towns (Italy) derived from SPOT-HRV Panchromatic image data (Figure 4.2).

Based on a consideration of the importance of the structural information, some systematic assessments of automatic feature detection routines have been made. These assessments are based on the systematic application of texture indices and classical morphological transformations, with a wide range of parameters (direction, size, kind of transformation) and

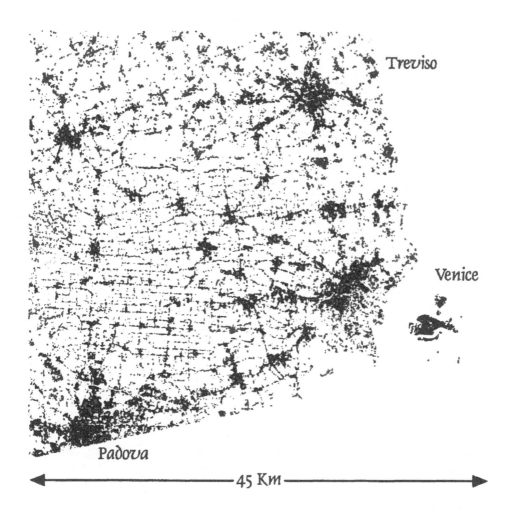

Figure 4.1: The built-up area in the 'dispersive city' of Venice-Padua-Treviso in the Veneto Region derived from a mixture of SPOT-HRV Panchromatic and Landsat TM image data.

62

Figure 4.2: The urban texture of the Bassano and Cittadella towns (Italy) derived from SPOT-HRV Panchromatic image data.

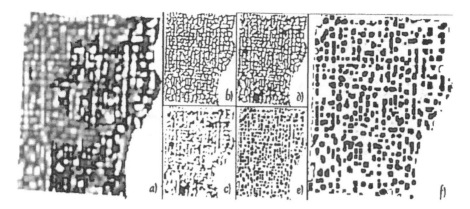

Figure 4.3: Automatic recognition of "pseudo-buildings" in SPOT-HRV image data: an application of the $r - h$ maximum morphological transformation.

the use of statistical measures of discriminatory power at selected test sites. Initial results proved very promising. The method has since been further improved and has been applied to a still wider range of geographical contexts — specifically the Veneto (Italy) Region, the Master Plan of the metropolitan area of Pescara (Italy), Durazzo (Albania), and San Salvador (El Salvador). The final release of the method is based on a fuzzy classification of textural layers calculated on the SPOT Panchromatic data. The thematic (urban vs. non-urban) classification accuracy is believed to be greater than 98%.

A more ambitious experiment of urban object recognition is provided by the SPOT-HRV Panchromatic data shown in Figure 4.3: the automatic detection of 'pseudo-building' and of other kinds of urban space (road and open spaces) is made through the application of a method known as $r - h$ maximum. This is a well known method of morphological segmentation, which uses digital, geodesic morphological transformations. Figure 4.3a shows the SPOT-HRV Panchromatic data for a representative urban area. These data are geometrically stretched by a factor of two (bilinear resampled) in order to detect structures one-pixel wide associated with the pseudo-building. Figures 4.3b–4.3e are radiometric threshold and morphological transformations applied to the $r - h$ transformed image. Figure 4.3f is the output of the segmentation process in which the black regions indicate candidate buildings, white areas represent the streets between them, and the grey tones indicate other urban open space. Clearly, the results are dependent on the spatial resolution of the sensor used and this technique can, in principle, be applied to the latest generation of very high resolution sensor instruments.

4.3.2 Segmentation of the Built-Up Areas

Another problem for urban remote sensing is how to segment urban areas which are homogeneous with respect to built form (e.g. type of housing) but which contain a range of site layouts. The spatial patterning of housing yields very important information for the architect-urbanist, as it provides an indicator of other attributes such as population density

Table 4.1: Morphological transition between 1970 and 1990 in the dispersive city of
Venice-Padua-Treviso (Italy). Units: hectares.

		1990					
		A	B	C	D	E	Total
	A	107,521	1,760	4,092	5,655	3,539	122,567
	B	-	3,090	867	409	110	4,475
1970	C	-	-	5,579	2,342	631	8,552
	D	-	-	-	5,843	3,062	8,905
	E	-	-	-	-	6,457	6,457
	Total	107,521	4,849	10,537	14,249	13,798	150,955

Key: A = background (no buildings); B = small accumulation ($L < 90m$); C = large
accumulation ($210m \leq L \leq 510m$); D = compact nucleus ($L \geq 510m$).

and composition, environmental impact, historical development and a range of cultural and
symbolic dimensions. Spatial pattern is a fundamental tool for the urban project, since it
invites hypothesis formulation. All this is well know to the urbanist, but studies of spatial patterning using automatic processing of satellite data are not yet numerous and have
always been experimental. The urbanist and the remote sensing specialist have much to
gain through collaboration on spatial pattern analysis. Some interesting examples of the
urban spatial pattern analysis have been made using texture indices and measures or local
heterogeneity, as well as morphological transformations and fractal analysis (Batty and Longley 1986, Frankhauser 1992). More recently, Eurostat has shown an interest in developing
related measures of the 'urban morphological zone' (Eurostat 1994, see also Weber, this
volume).

If we apply the same data processing techniques to images acquired on at least two
different dates, it is possible to describe the dynamics of an urban area. This approach not
only allows us to establish changes in number of urban pixels between the two images, but
also to record the transitions between classes in a 'morphological transition table' such as
Table 4.1.

4.4   Beyond Classical Thematic Information

In the preceding examples, structural information was used in the identification and description of urban areas. Questions like 'where are the urban areas?' and 'which type of urban
area is it?' were answered through thematic classification. But not all pertinent questions
may be reduced in this way: a set of 'softer' or 'fuzzy' questions also exists. They result in
images that are not reducible to maps of discrete land cover and land use parcels, but they
are equally important to the urbanist. They include treatments that emphasize some aspects
of the image, or some isolated structural characteristic that could be used as a basis for the
interpretation of process.

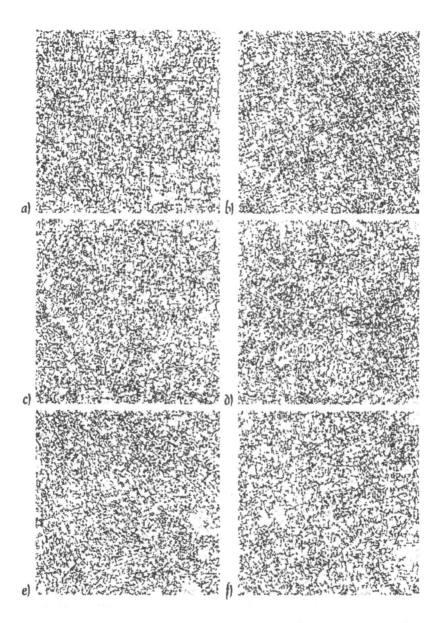

Figure 4.4: Study of the structural characteristics of the rural landscape through mathematical morphology analysis ('top-hat' transformation) of Landsat TM data: a) and b) — different orientations of the ancient Roman Centuriazio; c) and d) — spatial organization around the high and low parts of the Brenta River; e) and f) — spatial organization around the ancient Postumia road and the new lines of communication.

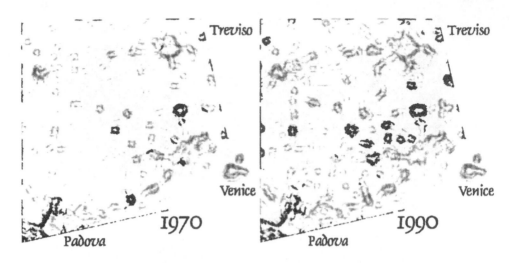

Figure 4.5: The evolution of the 'urban wall' in the dispersive city of Venice-Padua-Treviso in the
Veneto Region, Italy between 1970 and 1990.

The output of fuzzy processing is not a complete semantically discretized thematic image, but an emphasized one, through local and convolution transformations or mixing of
spectral and structural information. Figure 4.4, for example, shows morphological transformations applied to a Landsat TM image, which results in a map of territorial 'lines' or
structural alignment. The result is a skeleton of an inverse 'top-hat' morphological transformation applied to the near infra-red channel of a Landsat TM image, resampled to $15m$ by
means of bilinear convolution. The lines indicate features in the original image that exhibit
a localized minimum in reflectance. This can be most readily visualized as a V-shape in the
reflectance 'surface' expressed by the image. The features picked out in this way include
drainage channels, canals and roads.

Figure 4.5 takes this process one stage further by illustrating the detection of changes
in the density of the dispersed city region of Venice-Padue-Treviso over a 20 year period[3].
This region can be perceived as a 'diffusive town' or as 'urbanized country': its evolution
cannot be described simply by growth around a single nucleus, but as metamorphosis and
morphogenesis of a field of structures. The notion of an urban limit therefore becomes
intrinsically problematic, but could be constructed using a gradient transformation applied
to urban density to detect the 'urban walls'. The urban texture in Figure 4.5 does not belong
to a semantic thematic class, but is the outcome of a hybrid procedure. The outcome of this
procedure is a product that presents fundamental information for urban analysis, and which
is neither conventional continuous nor discrete classification. Is not possible to recover this
kind of information using traditional classification systems, because of the difficulties in
distinguishing object from the urban background at the available sensor resolution.

---

[3]DAEST-MURST research project "La città diffusa" (1992).

4.5 References

Aymonino, C., 1977, *Lo Studio dei Fenomeni Urbani* (Rome: Officina Edizioni).

Batty, M., and Longley, P. A., 1986, The fractal simulation of urban structure. *Environment and Planning A*, **18**, 1143–1179.

De Carlo, G. C., 1966, *La Pianficazone Territoriale Urbanistica nell'area Milanese* (Venice: Marsilio).

Eurostat, 1994, *Pilot Project Delimitation of Urban Agglomerations by Remote Sensing: Technical Report* (Luxembourg: Office for Official Publications of the European Communities).

Flouzat, G., Guichou, C., and Merghoub, Y., 1984, Recherche de fonctions texturantes et cartographic automatique de l'occupation des terres. *Espace Geographique*, **3**.

Frankhauser, P., 1992, Fractal properties of settlement structures, In *First International Seminar on Structural Morphology*, Montpellier, France.

Ginzburg, C., 1981, *Indagine su Piero* (Torino: Einaudi).

Gong, P., and Howarth, P. J., 1990, An assessment of some factors influencing multispectral land-cover classification. *Photogrammetric Engineering and Remote Sensing*, **56**, 597–603.

Haralick, R., and Saphiro, L., 1985, Image segmentation techniques. *Computer Vision, Graphics, and Image Processing*, **29**.

Harvey, D., 1969, *Explanation in Geography* (London: Edward Arnold).

Marceau, D. J., Howarth, P. J., Dubois, J. M. M., and Gratton, D. J., 1990, Evaluation of the gray-level co-occurrence matrix-method for land- cover classification using SPOT imagery. *IEEE Transactions on Geoscience and Remote Sensing*, **28**, 513–519.

Whitehand, J. W. R. (editor), 1981, *The Urban Landscape: Historical Development and Management* (London: Academic Press).

Zucker, S., 1976, Region growing: childhood and adolescence. *Computer Graphics and Image Processing*, **5**.

PART III

FROM URBAN LAND COVER TO LAND USE

CHAPTER 5

Modified Maximum-Likelihood Classification Algorithms and their Application to Urban Remote Sensing

Victor Mesev, Ben Gorte and Paul A. Longley

5.1 Introduction

Urban areas present one of the more challenging problems in image classification. Their typically complex mixture and erratic spatial arrangement of artificial and natural land cover types ensure that the reflectance values associated with individual pixels are often the result of the interaction with more than one surface material (Forster 1985). Consequently, per-pixel classifications of urban areas have been traditionally constrained to somewhat broad distinctions of urban/non-urban land cover, built/non-built and, at best, categorical building types. For practical urban monitoring, management and planning activities, however, much more detail is needed on the morphology and functional land use of settlements. To improve upon the detail and accuracy of per-pixel classification, research was at first targeted on how groups of neighbouring pixels can reveal the inherent textural (Webster *et al.* 1991), contextual (Harris 1985), and spatial (Barnsley *et al.* 1993) properties and patterns in urban images. Recent research has examined how neural networks can be trained to infer urban land cover; this includes an optimized neural-net technique which processes both spectral signatures and textural features based on grey-level histograms (Dreyer 1993). Notable contributions have even attempted to determine the composition of discrete pixels using, for instance, fuzzy sets (Fisher and Pathirana 1990) or assuming linear relationships between pixel values and land cover proportions (Forster 1985). Most of these and other techniques, however, have had somewhat variable degrees of success in producing classifications that are more accurate than simple per-pixel categorization. Indeed, bearing in mind the addi-

tional calculations and information needed to perform neighbourhood or sub-pixel classifications, it is instructive to note that similar levels of accuracy have been produced from
standard per-pixel methods, where these are adequately trained by reliable class signatures.
Equally favourable classification results may also be achieved if the per-pixel algorithm is
itself modified to take into account additional, non-spectral information on the structure and
characteristics of urban areas. This chapter will outline ways in which such additional data,
which are most conveniently handled within Geographical Information Systems (GIS), can
be used to modify the standard maximum likelihood (ML) per-pixel classifier. In this way,
improvements in detail and accuracy can then be routinely derived from ML classifications
which have been supplemented by socio-economic and cadastral data within an integrated
remote sensing/GIS (RS/GIS) framework. Here we explore one such integration methodology, but more comprehensive discussions may be found in Star *et al.* (1991) and, more
specifically pertaining to urban studies, in Langford *et al.* (1991) and Mesev *et al.* (1996).

Our methodology hinges on the ability of ancillary spatial data to identify and update the
a priori probabilities of the ML algorithm. This involves the use of these data first to stratify
images of urban areas according to spatial and contextual rules, and then to estimate the area
of the urban classes within each stratum. These area estimates are then either directly inserted into the ML classifier as prior probabilities, or are used as part of an iterative process
for creating and updating ML *a posteriori* probabilities. The use of ancillary spatial data to
stratify feature space is a progression of work pioneered by Strahler (1980), who demonstrated improvements in classification accuracy of natural vegetation in Doggett Creek, Cal-
ifornia. Since then, the technique has been further elaborated by Haralick and Fu (1983),
and supported by other work on the physical landscape. The most notable contributions
include Skidmore and Turner (1988), who extracted class probabilities from grey-level frequency histograms, and Maselli *et al.* (1992), who formalized a parametric/non-parametric
link between the ML discriminant function and prior probabilities. Apart from research by
Barnsley *et al.* (1989), Mesev *et al.* (1996) and Mesev (1998), work in the urban sphere
seems to have been largely neglected, undoubtedly because of the spectral heterogeneity of
urban land use categories. It is exactly when classes are closely related, however, that prior
probability estimates have the greatest effect (Mather 1985, Mesev 1998).

In this Chapter, we give a comprehensive account of how prior probabilities can be used
to modify the ML classifier and lead to improved areal estimates of spectrally similar classes
in the urban sphere. The Chapter also explores the role of ancillary spatial data both to constrain and to derive prior probabilities through stratification and re-iterative areal estimation,
respectively. We then demonstrate our techniques with empirical applications dealing with
urban land cover and land use classifications of settlements in the United Kingdom and the
Netherlands. The United Kingdom example uses population census data to modify the *a
priori* probabilities of four dwelling-type categories — detached, semi-detached, terrace,
and apartment blocks — while the Netherlands application accesses land-use planning information to aid in the prior-posterior iterative improvement of residential and industrial categories. The Chapter concludes with a discussion of the operational efficiency and practical
expediency of per-pixel ML classifications based on externally modified prior probabilities.
We also provide a note on how our work has contributed to a closer integration of remotely-
sensed and GIS data, and how spatial analysis and various applications of urban monitoring
may benefit from these closer links.

5.2   Modified Maximum-Likelihood Classification

As a parametric classifier, the ML algorithm relies on each training sample being represented by a Gaussian probability density function, completely described by the mean vector and variance-covariance matrix, using all available spectral bands. Given these parameters, it is possible to compute the statistical probability of a pixel vector being a member of each spectral class (Thomas *et al.* 1987). The goal is to assign the most likely class w_j, from a set of N classes, $w_1, \ldots, w_N$, to any feature vector x in the image. A feature vector x is the vector $(x_1, x_2, \ldots, x_M)$, composed of pixel values in M features (in most cases, spectral bands). The most likely class w_j for a given feature vector x is the one with the highest posterior probability $Pr(w_j|x)$. Therefore, all $Pr(w_j|x)$, $j \in [1 \ldots N]$ are calculated, and the w_j with the highest value is selected. The calculation of $Pr(w_j|x)$ is based on Bayes' Theorem,

$$Pr(w_j|x) = \frac{Pr(x|w_j) \times Pr(w_j)}{Pr(x)} \qquad (5.1)$$

On the left-hand side is the *a posteriori* probability that a pixel with feature vector x should be classified as belonging to class w_j. The right-hand side is based on Bayes' Theorem, where $Pr(x|w_j)$ is the conditional probability that some feature vector x occurs in a given class, in other words, the probability density of w_j as a function of x. Supervised classification algorithms, such as ML, derive this information from training samples. Often, this is done parametrically by assuming normal class probability densities and estimating the mean vector and covariance matrix. Alternatively, it is possible to use non-parametric methods, such as k-Nearest Neighbour (k-NN). The 'standard' k-NN methods directly implement a decision function based on the number of training pixels per class proportional to the prior probability (Fukunaga and Hummels 1987, Therrien 1989). This is the *a priori* probability of the occurrence of w_j irrespective of its feature vector, and as such is open to estimation by *a priori* knowledge external to the remotely-sensed image. External prior knowledge will typically include information on the spatial distribution and relative areas covered by each class in the study scene and is most readily generated from ancillary spatial data. It follows that the accuracy of class priors is, at best, equal to the quality of this *a priori* knowledge. In image classification terms, prior probabilities can be visualized as a means of shifting decision boundaries to produce larger volumes in M-dimensional feature space for classes that are expected to be large, and smaller volumes for classes that are expected to be small, in terms of numbers of pixels. The denominator in Equation (5.1), $Pr(x)$, is the unconditional probability density which is used to normalize the numerator such that

$$Pr(x) = \sum_{j=1}^{N} Pr(x|w_j) \times Pr(w_j) \qquad (5.2)$$

Typically, ML classifiers assume prior probabilities to be equal and assign each $Pr(w_j)$ a value of 1.0. However, variations in prior probabilities can be an important remedy for the problem of spectrally overlapping classes. If a feature vector x has probability density values that are significantly different from zero for several classes, it is not inconceivable for that pixel to belong to any of these classes. When selecting a class solely on the basis of its spectral characteristics, a large probability of error inevitably results. The use of appropriate prior probabilities, based on reliable supplementary information, is one way to reduce this error in class assignments. Moreover, it would seem intuitively more sensible to suggest that some classes are more likely to occur than others. At this stage of the discussion, it is important to differentiate between 'global', 'individual', and 'local' priors. Many proprietary software packages allow the use of global prior probabilities, where the user is expected to estimate them simply using information on the anticipated (relative) class areas. The improvement in classification is often limited. At the other extreme, each pixel can be influenced by a vector of prior probabilities that is valid only for that individual pixel. This is meaningless, however, because if the correct prior probabilities for each individual pixel were known beforehand, the classification would not be necessary! Given these problems, a compromise somewhere between global and individual priors can be derived, first by subdividing the image into strata (or segments) according to the ancillary spatial data, and then by finding the local prior probability vector for each stratum. In both of our examples, GIS data are used to stratify satellite images according to some contextual rules. In the first, spatially-distributed housing data from the 1991 UK Census of Population are used to assist in the selection of training samples for different dwelling types within, and the post-classification of, the residential stratum. From there, local prior probabilities are derived from the census data and applied to ML classification. The second example, taken from the Netherlands, uses planning data to aid training sample selection, once again. This time, however, the local prior probabilities are constantly updated by means of an iterative process using posterior probabilities. Instead of having to enter the priors at the beginning of the process, the user receives them at the end. In each case, we suggest that a systematic RS/GIS strategy should coordinate the flow and coupling of GIS data within image classification procedures.

5.2.1   Modification of Prior Probabilities

For our first example, prior probabilities will be modified using a hierarchical stratification scheme based on data from the United Kingdom Census of Population. The stratification allows the Census data to assist in the selection and hierarchical partitioning of spatial features from a satellite image. For prior probabilities to function most efficiently they need to operate within inclusive feature space and derive mutually-exclusive classes. In other words, for the classification of mutually-exclusive residential dwelling classes, an image must only be composed of residential feature space. Census data have already been shown to be capable of generating pseudo-surfaces of urban representations, especially residential surfaces (Martin and Bracken 1991), from which such stratification is possible. These surfaces have been used by Mesev (1998) to enhance per-pixel classifications through training sample selection and post-classification sorting. The result is that satellite images have been routinely segmented into 'urban' and 'non-urban', as well as 'residential urban' and 'non-residential urban' classes (Mesev *et al.* 1995). Using the 'residential urban' category we show how prior probabilities of the surrogate residential density categories 'detached', 'semi-detached', 'terrace', and 'apartment' blocks, may be generated by Census data and inserted into the ML classifier.

Consider z_k as the Census variable 'residential building type' (where k : 1='detached', 2='semi-detached', 3='terrace', and 4='apartments'). When stratified into exclusively residential feature space, the four classes will have A pixels with feature values x_i, where x_1, ..., x_A are not necessarily mutually-exclusive. The objective is to find the probability that a pixel, selected at random from the 'residential' stratum of the image, will be a member of a spectral class w_j (where j : 1='detached', 2='semi-detached', 3='terrace', 4='apartments'), given its density vector of observed measurements $\mathbf{x}$, in m-dimensional feature space *and* that it belongs to ancillary class z_k, described as:

$$Pr(w_j|\mathbf{x}, z_k) \qquad (5.3)$$

It is also assumed that the effects of z_k are external to the original generation of the mean vector and covariance matrix of w_j. As a result, the likelihood function $Pr(w_j|\mathbf{x})$ is unaltered by the introduction of z_k, but is simply modified by the conditional probability:

$$Pr(w_j|z_k) \qquad (5.4)$$

This is a process of identifying the association between spectral class w_j with census variable z_k. For example, the spectral class labelled as 'low density residential' would be directly associated with a conditional probability of the census variable, 'detached dwellings'. In effect, w_1 is weighted by the probability of z_1, producing the prior probability of $Pr(w_1)$. In our example we assume that the prior probabilities of each of the four dwelling types exist in inclusive m-dimensional feature space, so that $Pr(w_1) + Pr(w_2) + Pr(w_3) + Pr(w_4) = 1.0$. The probability densities $d_{i1} = Pr(x_i|w_1)$, $d_{i2} = Pr(x_i|w_2)$, $d_{i3} = Pr(x_i|w_3)$, $d_{i4} = Pr(x_i|w_4)$, are known for each pixel. Let l_{i1} be the shorthand for the posterior probability $Pr(w_1|x_i, z_1)$ that pixel i belongs to class w_1, and let p_j be the shorthand for the prior probabilities. The Bayesian modified ML is now represented as

$$l_{i1} = \frac{d_{i1}p_1}{d_{i1}p_1 + d_{i2}p_2 + d_{i3}p_3 + d_{i4}p_4} \qquad (5.5)$$

Likewise, $l_{i2} = Pr(w_2|x_i, z_2), l_{i3} = Pr(w_3|x_i, z_3)$ and $l_{i4} = Pr(w_4|x_i, z_4)$ may also be found, and of course, the sum of the four posterior probabilities equals 1.0,

$$l_{ij} = \frac{d_{ij}p_j}{\sum\limits_{j=1}^{4} d_{ij}p_j} \qquad (5.6)$$

5.2.2 Empirical Example 1

We will now examine an empirical application of the Bayesian-modified ML classifier to the settlement of Norwich in eastern England; others may be found in Longley and Mesev (1997) and Mesev (1998). The aim is to produce a modified ML classification into the four residential dwelling types outlined above, from a Landsat 5 TM image taken on the 15th July 1989, using primarily the Unix-based ERDAS (*Imagine* 8.2) image processing software (ERDAS 1995) as well as some purpose-written programs. Before the modified ML

Table 5.1: Classification results using equal and unequal prior probabilities.

Dwelling Type	Census		Equal Priors			Unequal Priors		
	EDs	%Area	Pixels	%Area	Error	Pixels	%Area	Error
Detached	37364	42.40	12486	38.75	-3.65	13910	43.17	+0.77
Semi-Detached	26675	30.27	10311	32.00	+1.73	9160	28.43	-1.84
Terraced	21088	23.93	8030	24.92	+0.99	7440	23.09	-0.84
Apartments	2987	3.39	1395	4.33	+0.94	1712	5.31	+1.92
Totals	88 114	100.00	32 222	100.00	7.33†	32 222	100.00	5.37†

† Total error for each dwelling type measured as the difference in area between the census and equal/unequal priors, expressed in absolute terms. ED = Enumeration District.

classifier is implemented, a series of hierarchical segmentations is carried out to partition each image and generate the 'residential' stratum from which the four dwelling types are ultimately derived. The first segmentations are based on standard unsupervised classifications, using ERDAS, and produce generalized 'urban' and 'non-urban' strata, from which the 'urban' stratum is subdivided into 'built-up' urban and 'non-built-up' urban. Housing data from the 1991 UK Census of Population are then used to help further partition the 'built-up' stratum into 'residential' built-up and 'non-residential' built-up (the complete hierarchical structure can be seen in Longley and Mesev, Chapter 9, this volume). As already mentioned, housing information are used here as input data for a GIS surface model, derived from (Martin and Bracken 1991), and applied to the selection of class training samples and post-classification sorting. The discrepancy in the dates between when the satellite images were taken and the 1991 Census was unavoidable, although this was not a particularly rapid period of house building in England.

The 'residential' stratum is eventually exposed to the modified ML classifier. The prior probabilities $Pr(w_j)$ for each dwelling type category are essentially their area estimates, and are calculated with reference to the proportion of households that fall within each dwelling type (Figure 5.1). These statistics are extracted directly from the Census, then normalized to create a probability distribution, and finally transformed into a usable form which takes into account the relative size ratios. This transformation helps to preserve the relative areal proportions of each dwelling type, where for instance 'detached' dwellings occupy larger areas than 'terrace' dwellings. Using stereoscopic aerial photographs, 20 samples of dwelling type sizes are generated and average relative size ratios between dwelling types constructed. The ratios stabilized at 1 detached dwelling to 1.5 semi-detached, 1 detached to 2.25 terrace, and 1 detached to 10 apartments. Although these are approximations, they are still more realistic than assuming absolute, 1:1 linear relationships.

The results of the thematic classifications of the four dwelling density types based on maximum *a posteriori* probabilities are shown graphically in Figure 5.2, together with area estimates in Table 5.1. The differences between equal and unequal prior probability classification are visually apparent in Figure 5.2. Many parts of Norwich, especially the north areas, have been classified with contrasting dwelling types. Table 5.1 further quantifies the

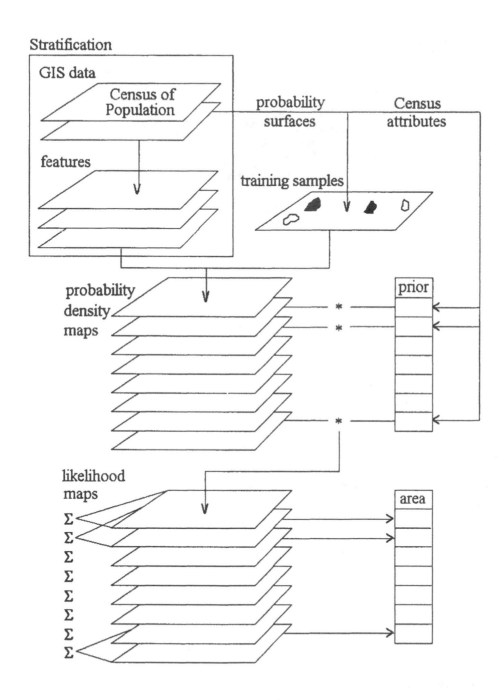

Figure 5.1: Insertion of prior probabilities in the maximum likelihood classifier.

78

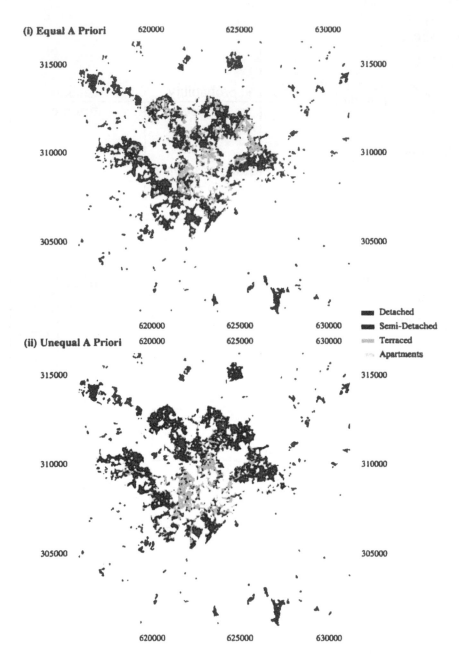

Figure 5.2: ML classifications of Norwich, U.K., using (i) equal *a priori*, and (ii) unequal *a priori* probabilities.

success of ML classifications using adjusted, or unequal, prior probabilities over classifications employing equal prior probabilities. Over all classes, the area estimates produced from the Bayesian modified-ML classifier are closer to those derived from the size-ratio transformed Census figures. The best results are obtained from the detached category, perhaps because of its larger area on the ground and hence the lower degree of spectral mixing. The worst is, understandably, apartments, where the size ratio may not have been truly representative. Further accuracy assessments for other settlements, including one with a detailed site-specific evaluation, can be found in Mesev (1995). All indicate slight to moderate improvements.

So far, we have established how GIS-informed changes to *a priori* class membership probabilities may be used to alter the standard ML classifier and hence improve classifications. The next section further assesses the benefits of using GIS data for calculating *a priori* probabilities. It also demonstrates how an innovative, iterative process can take advantage of maximum *a posteriori* probabilities to modify further *a priori* probabilities.

5.2.3   Iterative Calculation of Prior Probabilities

Area estimates are routinely generated from standard ML classifications simply by referring to the class histogram and counting the number of pixels in each class. Because ML classifiers are commonly biased in favour of some classes, however, these estimates are unreliable indicators of class sizes (Conese and Maselli 1992). Nevertheless, class area estimates are still important indicators of classification accuracy and a process whereby these estimates contribute towards improved classification will now be outlined. This second modification of the standard ML classification procedure, therefore, relies not only on the use of maximum *a posteriori* probabilities to label spectral clusters but also to obtain class area estimates. Using these estimates, an updated set of prior probabilities is calculated and the classification repeated. Thus, the process becomes iterative and eventually converges to statistically correct area estimates. As in the previous modification, the process is most effective when applied to stratified images which have been produced in association with additional spatial data (Figure 5.3). The approach relies on maintaining the entire vector of posterior probabilities, as well as making a maximum *a posteriori* probability decision for each pixel (conventional ML classification). The sum of these vectors of posterior probabilities yields an estimate for the vector $(A_1, \ldots, A_N)$ of the area per class. The areas are measured in pixels, and by normalizing the areas we obtain the vector of prior probabilities. We will now detail the iterative prior probability methodology and provide a simple example. The full mathematical calculations appear in the Appendix to this Chapter.

Suppose that 100 image pixels have the same feature vector x and that the posterior probability $Pr(w_j|\mathbf{x}) = 0.77$. This should be interpreted as 77 out of those 100 pixels will be expected to belong to class w_j; the other 23 will belong to other classes. Unfortunately, the ML classifier is unable to report which 23. We can state, however, that out of those 100 pixels, 77 will contribute to the area of class w_j. Moreover, by using the other components of the posterior probability vector, we can find out the contributions of the 100 pixels to the areas of the other classes. The example can be made independent of the number 100 by saying that each pixel contributes 0.77 to class w_j and a cumulative 0.23 to other classes. In contrast to the first modification of the ML classifier, posterior probability vectors are now

80

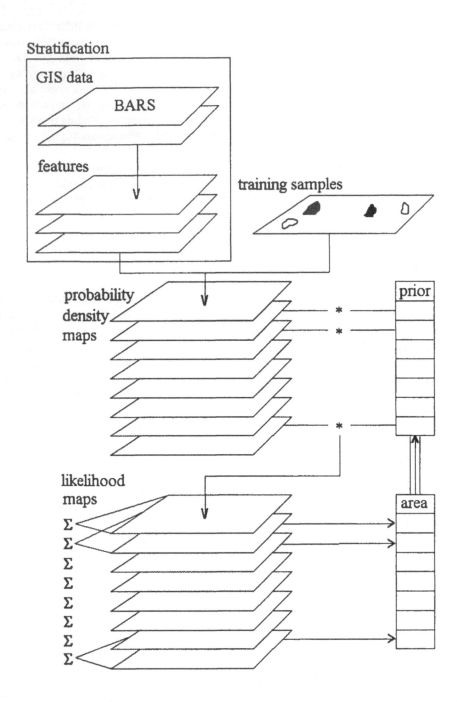

Figure 5.3: Iterative calculation of prior probabilities.

used to calculate the prior probabilities of the next classification. This is done by converting the posterior probability vectors into class areas and normalizing. If we guess the initial prior probability correctly, the ones we find at the end of this process should be the same. If, as is the more usual case, they are wrong, the final ones will be different but, importantly, they will have improved and the whole process can be repeated using successively 'new' priors (in the true sense of Bayes' Theorem). The process becomes iterative, as illustrated by Figure 5.3. The question of whether it converges to the true 'prior' probabilities is addressed in the Appendix to this Chapter.

To illustrate this iterative process, consider the simple situation in which there are only two classes (w_1 and w_2) and A pixels having feature vectors $x_1 \dots x_A$ with known probability densities d_{i1} and d_{i2} for each pixel. In effect, this is a collection D of A probability density vectors (noting that D is not a set, since duplicates may occur),

$$\mathbf{D} = \left[\begin{array}{c} d_{11} \\ d_{12} \end{array} \right], \left[\begin{array}{c} d_{21} \\ d_{22} \end{array} \right], \dots, \left[\begin{array}{c} d_{A1} \\ d_{A2} \end{array} \right] \qquad (5.7)$$

After faithfully applying Bayes' rule A (or $2 \times A$) times, we can obtain a collection L with A posterior probability vectors,

$$\mathbf{L} = \left[\begin{array}{c} l_{11} \\ l_{12} \end{array} \right], \left[\begin{array}{c} l_{21} \\ l_{22} \end{array} \right], \dots, \left[\begin{array}{c} l_{A1} \\ l_{A2} \end{array} \right] \qquad (5.8)$$

The main postulate of this iterative modification to the ML classifier is that the sum of the vectors in L should equal the vector $[A_1, A_2]$, which is the total areas covered by w_1 and w_2, respectively,

$$\sum_{i=1}^{A} l_{i1} = \sum_{i=1}^{A} \frac{d_{i1} p_1}{d_{i1} p_1 + d_{i2} p_2} = A p_1 \qquad (5.9)$$

and

$$\sum_{i=1}^{A} l_{i2} = \sum_{i=1}^{A} \frac{d_{i2} p_2}{d_{i1} p_1 + d_{i2} p_2} = A p_2 \qquad (5.10)$$

(Note that these equations are equivalent, since $l_{i1} + l_{i2}$ and $p_1 + p_2 = 1$). Remember that the priors, p_1 and p_2, are presumed to be relative areas, such that $p_1 = \frac{A_1}{A}$ and $p_2 = \frac{A_2}{A}$. Furthermore, it is our contention that, if the prior probabilities p_1 and p_2 are unknown, they can be obtained by solving Equation 5.9. The solution depends on the matrix D only. A detailed analysis of the mathematical operations that surround the calculation of the D matrix is given in the Appendix to this Chapter.

Table 5.2: Relative areas (%) per stratum.

Stratum	Area %	Water	Grass	Agric.	Forest	Resident.	Green	Indust.	Bare
Residential	6.3	0.0	0.0	0.2	0.0	99.1	0.1	0.1	0.5
Industrial	2.6	0.0	0.0	4.0	0.0	1.9	0.1	91.8	2.2
Green	26.8	0.0	0.0	0.0	0.0	9.5	84.0	6.5	0.1
Bare	0.5	0.0	0.0	1.4	0.0	0.0	0.0	0.2	98.3
Other	33.1	24.4	41.4	12.6	1.6	0.4	0.1	0.8	18.7
Residential/Industrial	1.1	0.0	0.0	0.1	0.5	94.7	0.3	4.3	0.1
Residential/Green	3.0	0.0	0.0	0.2	0.0	72.6	18.5	8.6	0.1
Residential/Bare	0.3	0.0	0.1	14.5	0.0	25.2	0.7	49.5	10.0
Residential/Other	2.3	0.0	11.8	5.2	1.6	66.4	0.1	0.3	14.5
Industrial/Green	1.5	0.0	0.0	0.1	0.0	55.4	19.1	25.4	0.1
Industrial/Bare	0.5	0.0	0.0	11.1	0.0	5.4	0.2	33.4	50.0
Industrial/Other	3.4	10.6	0.9	6.0	0.1	42.5	2.0	21.5	16.5
Green/Bare	0.3	0.0	23.4	0.2	0.0	4.7	3.2	47.1	21.4
Green/Other	15.4	0.0	11.2	9.9	0.1	7.5	43.3	6.9	21.2
Bare/Other	0.7	16.0	2.5	7.5	0.0	6.7	0.1	50.1	17.2

5.2.4 Empirical Example 2

The ML modification using iterative priors has been tested in a number of cases using real image data as well as synthetic data, and so far has produced encouraging results. One of these empirical applications concerns Westland, in the province of Zuid Holland in the Netherlands, an area characterized by greenhouses which cover approximately 35% of the land. The objective is to assess the degree to which rapidly expanding residential and industrial land is impinging on agricultural fields and grasslands. Other land cover types discernible in the available Landsat TM image of this scene include a part of the Nieuwe Maas–Nieuwe Waterweg, connecting Rotterdam with the North Sea, as well as some forested areas. The initial, conventional ML classification produced eight classes, namely 'greenhouses', 'residential', 'industrial', 'agriculture', 'grass', 'bare soil', 'forest', and 'water'. Of the first three 'urban' classes, 'residential' and 'industrial' were not very homogeneous and largely overlapped each other in multispectral feature space. Not surprisingly, 'greenhouses' were found throughout the feature space, simply because their detected reflectance was very much dependent on the orientation of the glass roofs with respect to the sun angle (azimuth and elevation). The 'agricultural' classes were also spectrally multi-modal, although 'grass', 'forest' and 'water' were all classified successfully using a non-parametric (k-NN) algorithm. Various instances of 'bare soil' were identified in unused (vacant) agricultural fields, building sites, as well as parts of the coastal dunes and beaches. More importantly, this class exhibited a similar spectral reflectance to sections of the 'residential' and 'industrial' categories.

In the context of our drive towards the integration of GIS data into image classification, additional information was again used to perform a spatial stratification of the area. Over 50 functional land use categories were available from BARS (Basisbestand Ruimtelijke Structuren), a data set compiled by the Dutch planning agency, RPD (Rijks Planologische Dienst). Temporal inconsistencies of 2–4 years between the compilation of the BARS data set and the acquisition of the Landsat image were unavoidable. The high generalization of BARS allowed five classes to be constructed, these being 'residential', 'industrial', 'greenhouse', 'bare soil' and 'rest' (other) (Figure 5.4), and constituted the first five strata. The other ten strata were derived from buffer zones generated around these land uses, taking into account the likelihood that new residential buildings will be constructed next to existing ones. Thus, buffer zones of a few hundred metres were created to represent 'grass near to residential' areas, and 'industrial near to residential' areas (Figure 5.4).

While performing the iterative process, the relative areas of the eight image classes were calculated for each stratum. The result after ten iterations is shown in Table 5.3. Convergence was reached in the sense that during the tenth iteration none of the probabilities changed more than 0.05%. Using the values in Table 5.2 (shown in bold) as prior probabilities, spatially distributed according to the stratification, an ML classification was performed.

The accuracy and reliability of the results generated from the modified ML classification were then evaluated using sets of 200 samples per class, which were taken from all over the image. From Table 5.3, both the residential category and the category representing the rest of the image increased in accuracy and reliability. An accuracy of 97% for the residential class is very high and an important vindication of the classification methodology. Conversely, the industrial class at best maintained the same level of accuracy as the 'standard' k-NN

84

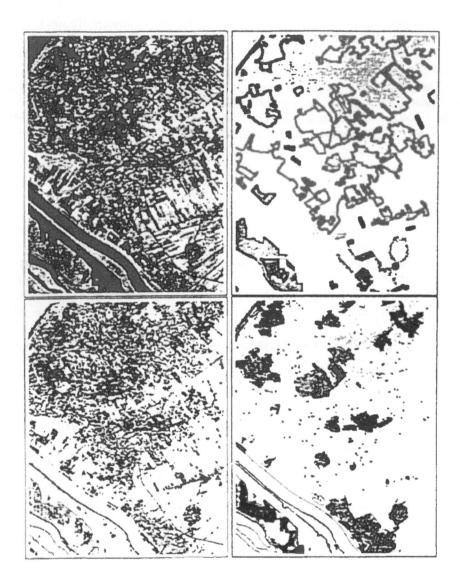

Figure 5.4: Results of iterative ML classification scheme. Upper left: Landsat-TM image (band 4); upper right: definition of strata; lower left: standard, non-parametric classification; lower right: final classification. In the classifications, three classes are identified: industrial (dark), residential (medium) and greenhouse (light); other classes are left blank. The vector coastline was added after classification.

Table 5.3: Classification accuracies and reliabilities.

Class	Standard		Final	
	Accuracy	Reliability	Accuracy	Reliability
Residential	0.88	0.66	0.97	0.78
Industrial	0.56	0.58	0.55	0.88
Green	0.41	0.70	0.85	0.76

classification. This was due in part to some systematic misclassification, for example, the entire beach was classified as industrial instead of bare soil. Nevertheless, the reliability of this class increased: 88% of the pixels classified as industrial belong to that class, while the other 12% can be accounted for by errors in the classification of land adjacent to the coast.

5.3 Discussion and Conclusions

Per-pixel image classification is an established part of remote sensing methodology. However, the inherent spatial heterogeneity of built surfaces has tended to restrict detailed image classifications of urban areas. In this Chapter, we have attempted to disentangle the complex spectral patterns associated with urban areas by contributing to the refinement of the ML per-pixel classifier, and through links with ancillary spatial (GIS) data. Indeed, it is precisely because we are dealing with spectrally overlapping classes that we have advocated two contributions to per-pixel classification, both based on and adhering to the probabilistic, and hence flexible, nature of the standard ML discriminant function. By careful modifications to prior probabilities, either directly using ancillary data or by an iterative process involving posterior probabilities, we can begin to pry open urban spectral classes. In each case, ancillary spatial data and GIS operations are seen as essential for the efficient implementation of these modifications, not only to assist in the calculation of prior probabilities, but also to support the stratification of the image without which prior probabilities could not function.

Although the ultimate goal of our work is to produce more accurate classifications, the two methodologies we have outlined also contribute to the continued integration of data and technology between remote sensing and GIS. The integration issues we have addressed are those relating to information exchange, positional integrity, and, most importantly, classification compatibility. Spatially-distributed data, in the form of a census surface and land use planning map, have been incorporated into the image classification process to select and label class training samples, as well as to calculate stratified prior probabilities. The final classified products have been further compared with and assessed against, the initial census surface and land use planning map. Integration may be extended still further if the information derived from our modified ML classifications were used directly in a GIS. For instance, the posterior probabilities of per-pixel class membership might be stored and updated in a GIS database to await interrogation. The database might be capable of resolving queries pertaining to land use change. For example, the user may wish to identify land use areas which may have changed from agriculture to industrial, over the past ten years, with probabilities of greater than 0.8. Another avenue of integration which might be pursued is that of linking classified data with spatial analysis. In Longley and Mesev (Chapter 9, this volume), two of the authors detail the application of spatial analysis techniques based on fractal geometry to understand further the structure and patterns of the Norwich study area using the classification developed in section 5.2.2. In particular, they examine how improvements in image classification of urban areas may be exploited by fractal models and enable GIS to monitor and analyze the size and form of urban morphologies, as well as the cumulative and density profiles generated between the city centre and the city periphery.

5.4   Acknowledgements

The contributions made by Victor Mesev were partly funded by a Research Fellowship (number H53627501295) awarded by the UK Economic and Science Research Council (ESRC).

5.5   References

Barnsley, M. J., Barr, S. L., Hamid, A., Muller, J.-P., Sadler, G. J., and Shepherd, J. W., 1993, Analytical tools to monitor urban areas, In *Geographical Information Handling — Research and Applications*, edited by P. Mather (Chichester: John Wiley and Sons), pp. 147–184.

Barnsley, M. J., Sadler, G. J., and Shepherd, J. W., 1989, Integrating remotely sensed images and digital map data in the context of urban planning, In *Proceedings of the 15th Annual Conference of the Remote Sensing Society*, Remote Sensing Society (Nottingham: Remote Sensing Society), pp. 25–32.

Conese, C., and Maselli, F., 1992, Use of error matrices to improve area estimates with maximum likelihood classification procedure. *Remote Sensing of Environment*, **40**, 113–124.

Dreyer, P., 1993, Classification of land cover using optimized neural nets on SPOT data. *Photogrammetric Engineering and Remote Sensing*, **59**, 617–621.

ERDAS, 1995, *ERDAS Imagine (8.2): User's Manual*, Earth Resources Data Analysis Systems Inc., Atlanta, GA, USA.

Fisher, P. F., and Pathirana, C., 1990, The evaluation of fuzzy membership of land cover classes in the suburban zone. *Remote Sensing of Environment*, **34**, 121–132.

Forster, B. C., 1985, An examination of some problems and solutions in monitoring urban areas from satellite platforms. *International Journal of Remote Sensing*, **6**, 139–151.

Fukunaga, K., and Hummels, D., 1987, Bayes error estimation using Parzen and k-NN procedures. *IEEE Transactions PAMI*, **9**, 634–643.

Haralick, R. M., and Fu, K., 1983, Pattern recognition and classification, In *Manual of Remote Sensing*, edited by R. Colwell, second edition (Falls Church, VA., USA: American Society of Photogrammetry and Remote Sensing).

Harris, R., 1985, Contextual classification post-processing of Landsat data using a probabilistic relaxation model. *Internaitonal Journal of Remote Sensing*, 6, 847–866.

Langford, M., Maguire, D. J., and Unwin, D. J., 1991, The areal interpolation problem: estimating population using remote sensing in a GIS framework, In *Handling Geographical Information: Methodology and Potential Applications*, edited by I. Masser, and M. Blakemore (Harlow: Longman), pp. 55–77.

Longley, P. A., and Mesev, T. V., 1997, The use of diverse RS-GIS sources to measure and model urban morphology. *Geographical Systems*, 4, 15–18.

Martin, D. J., and Bracken, I., 1991, Techniques for modelling population-related raster databases. *Environment and Planning A*, 23, 1069–1075.

Maselli, F., Conese, C., Petkov, L., and Resti, R., 1992, Inclusion of prior probabilities derived from a non-parametric process into the maximum-likelihood classifier. *Photogrammetric Engineering and Remote Sensing*, 58, 201–207.

Mather, P. M., 1985, A computationally-efficient maximum likelihood classifier employing prior probabilities for remotely-sensed data. *International Journal of Remote Sensing*, 6, 369–376.

Mesev, T., 1998, The use of census data in urban image classification. *Photogrammetric Engineering and Remote Sensing*, 64, 431–438.

Mesev, T., Longley, P., Batty, M., and Xie, Y., 1995, Morphology from imagery: detecting and measuring the density of urban land use. *Environment and Planning A*, 27, 759–780.

Mesev, T. V., 1995, *Urban land use modelling from classified satellite imagery*, Ph.D. thesis, School of Geographical Sciences, University of Bristol.

Mesev, T. V., Longley, P. A., and Batty, M., 1996, RS/GIS and the morphology of urban settlements, In *Spatial Analysis: Modelling in a GIS Environment*, edited by P. A. Longley, and M. Batty (Cambridge: GeoInformation International), pp. 123–148.

Skidmore, A. K., and Turner, B. J., 1988, Forest mapping accuracies are improved using a supervised nonparametric classifier with SPOT data. *Photogrammetric Engineering and Remote Sensing*, 54, 1415–1421.

Star, J. L., Estes, J. E., and Davis, F., 1991, Improved integration of remote sensing and geographic information: a background to NCGIA initiative 12. *Photogrammetric Engineering and Remote Sensing*, 57, 643–645.

Strahler, A. H., 1980, The use of prior probabilities in maximum likelihood classification of remotely-sensed data. *Remote Sensing of Environment*, 10, 135–163.

Therrien, C. W., 1989, *Decisions, Estimation and Classification* (Chichester: John Wiley and Sons).

Thomas, I. L., Benning, V. M., and Ching, N. P., 1987, *Classification of Remotely-Sensed Images* (Bristol: IOP).

Webster, C. J., Oltof, W., and Berger, M., 1991, Exploring the discriminating power of texture statistics in interpreting a SPOT satellite image of Harare, Technical Report in GeoInformation Systems, Computing and Cartography 36, University of Wales Cardiff, Wales and South West Regional Research Laboratory.

Appendix

This Appendix expands upon the mathematical formulation of the iterative calculation of class areas. Assume three small examples of $\mathbf{D}$, $\mathbf{D}_1$, $\mathbf{D}_2$ and $\mathbf{D}_3$, given as

$$\mathbf{D}_1 = \begin{bmatrix} 4 & 2 & 2 & 1 & 1 & 1 \\ 1 & 1 & 1 & 1 & 2 & 3 \end{bmatrix},$$

$$\mathbf{D}_2 = \begin{bmatrix} 4 & 2 & 2 & 1 & 1 & 2 \\ 1 & 1 & 1 & 1 & 2 & 3 \end{bmatrix}, \tag{5.11}$$

$$\mathbf{D}_3 = \begin{bmatrix} 4 & 2 & 2 & 1 & 1 & 4 \\ 1 & 1 & 1 & 1 & 2 & 7 \end{bmatrix}.$$

(when dividing all of the numbers by 1000 it becomes easier to imagine them as probability densities, but it will have no effect on the results because of normalization in Equations 5.9 and 5.10). Since it may not be obvious to the reader why $\mathbf{D}_1$ gives a 'regular' solution $0 < p_1 < 1$, $\mathbf{D}_2$ does not give a regular solution, and $\mathbf{D}_3$ yields $p_1 = 1$, we discuss a number of further considerations, namely:

- What conditions $\mathbf{D}$ must satisfy to yield a unique solution?

- How can we find this solution?

- Does $\mathbf{D}$ meet the conditions if it consists of probability densities?

Conditions to D

If we concentrate on Equation 5.9 and try to solve for p_1, we must first re-write Equation 5.10 as

$$\Im(p_1) = \sum_{i=1}^{A} f(p_1) = 0 \tag{5.12}$$

with

$$f_i(p_i) = \frac{d_{i1} p_1}{d_{i1} p_1 + d_{i2} p_2} - p_i \tag{5.13}$$

Next, we observe that, because of the normalization in Equation 5.5, the posterior probabilities for each pixel depends solely on the ratio between the two densities, rather than on their absolute values. This implies that zeros and negative values are not permitted in $\mathbf{D}$, which makes sense for class probability densities. Therefore, instead of $\mathbf{D}$ we can use a collection $\mathbf{E}$

$$\mathbf{E} = \begin{bmatrix} e_1 \\ 1 \end{bmatrix}, \begin{bmatrix} e_2 \\ 1 \end{bmatrix}, \dots \begin{bmatrix} e_A \\ 1 \end{bmatrix} = \begin{bmatrix} \frac{d_{11}}{d_{12}} \\ 1 \end{bmatrix}, \begin{bmatrix} \frac{d_{21}}{d_{22}} \\ 1 \end{bmatrix}, \dots \begin{bmatrix} \frac{d_{q1}}{d_{q2}} \\ 1 \end{bmatrix} \tag{5.14}$$

This allows us to change Equation 5.5 into

$$f_i(p_i) = \frac{e_i p_1}{e_i p_1 + p_2} - p_1 \tag{5.15}$$

and, since $p_2 = 1 - p_1$,

$$f_i(p_i) = \frac{e_i p_i}{(e_i - 1)p_1 + 1} - p_1 \tag{5.16}$$

Since $f_i(0) = 0$ and $f_i(1) = 0$, it is clear that $p_1 = 0$ and $p_1 = 1$ are solutions of Equation 5.12. We will call these trivial solutions because they reduce the classification to a one-class 'problem'. Moreover, it is likely that there are solutions for which $p_1 < 0$ or $p_1 > 1$. We will not be concerned about them, since p_1 is a probability. The question is whether there is a unique solution for $0 < p_1 < 1$. Let us examine the f_i type of functions. For convenience, we simplify the notation slightly and write Equation 5.16 as

$$f_e(p) = \frac{ep}{(e - 1)p + 1} - p \tag{5.17}$$

The first, second and third derivatives f_e', f_e'' and f_e''' with respect to p will be needed. For convenience, $g_e(p)$ is defined as the denominator of the first term of f_e, i.e. $g_e(p) = (e - 1)p + 1$, and $g_e'(p) = e - 1$. In the interval of interest, $0 \le p \le 1$, $f_e(p)$ is continuous and $g_e(p) > 0$.

$$f_e(p) = \frac{ep}{g_e(p)} - p \tag{5.18}$$

$$
\begin{aligned}
f_e'(p) &= \frac{eg_e(p) - g_e'(p)ep}{g_e^2(p)} - 1 \\
&= \frac{e((e-1)p+1) - (e-1)ep}{g_e^2(p)} - 1 \\
&= \frac{e(ep-p+1) - e^2p + ep}{g_e^2(p)} - 1 \tag{5.19}\\
&= \frac{e^2p - ep + e - e^2p + ep}{g_e^2(p)} - 1 \\
&= \frac{e}{g^2(p)} - 1
\end{aligned}
$$

$$f_e''(p) = \frac{-2e(e-1)g_e(p)}{g_e^4(p)} = \frac{-2e(e-1)}{g_e^3(p)} \tag{5.20}$$

$$f_e'''(p) = \frac{2e(e-1)3g_e^2(p)(e-1)}{g_e^6(p)} = \frac{6e(e-1)^2}{g_e^4(p)} \tag{5.21}$$

We observe that in the range $0 < p < 1$:

- $f_e(0) = 0$ and $f_e(1) = 0$

- if $e > 1$, $f''_e(p) < 0$, so f_e is convex and $f_e(p) > 0 \, \forall p : 0 < p < 1$

- if $e < 1$, $f''_e(p) > 0$, so f_e is concave and $f_e(p) < 0 \, \forall p : 0 < p < 1$

- if $e = 1$, $g_1(p) \equiv 1$ and $f_1(p) \equiv 0$

Returning to the problem of solving $\Im(p) = 0$, where $\Im(p)$ is the sum of A functions $f_e(p)$, the question now is which combinations of f_e's give non-trivial solutions and which do not. The answer is given by the derivative of $\Im'(0)$ and $\Im'(1)$. If they have different signs, or if one equals zero, then the function is either entirely positive or entirely negative (within $0 < p < 1$). Only if both derivative values are positive will the function resemble the 'sum' and have a non-trivial solution. Another possibility for a non-trivial solution would be if both derivative values are negative. However, this will not happen because $\Im''(p) = \sum f''(p) > 0$ (i.e. as p increases, $\Im$ can change from convex to concave, but not the other way around). This also explains why there will never be more than one non-trivial solution. Consequently, for a non-trivial solution we need $\Im'(0) > 0$ and $\Im''(1) > 0$, where

$$
\begin{aligned}
\Im'(0) &= \sum_{i=1}^{A} f'_i(0) \\
&= \sum_{i=1}^{A} \left(\tfrac{e_i}{g_i^2(0)} - 1 \right) \\
&= \sum_{i=1}^{A} (e_i - 1) \\
&= \sum_{i=1}^{A} \left(\tfrac{d_{i1}}{d_{i2}} - 1 \right) \\
&= \sum_{i=1}^{A} \tfrac{d_{i1}}{d_{i2}} - A
\end{aligned}
\tag{5.22}
$$

and

$$
\begin{aligned}
\mathcal{G}'(1) &= \sum_{i=1}^{A} f_i'(1) \\
&= \sum_{i=1}^{A} \left(\frac{e_i}{g_i^2(1)} - 1 \right) \\
&= \sum_{i=1}^{A} \left(\frac{e_i}{e_i^2} - 1 \right) \\
&= \sum_{i=1}^{A} \left(\frac{1}{e_i} - 1 \right) \\
&= \sum_{i=1}^{A} \frac{d_{i2}}{d_{i1}} - A
\end{aligned}
\tag{5.23}
$$

must both be greater than 0. This leads to a result which is remarkable enough to be called a Lemma.

Lemma: Given a collection D of class probability densities d_{ci}, exactly one-trivial solution for the area estimates will be found if

$$
\sum_{i=1}^{A} \frac{d_{i1}}{d_{i2}} > A
\tag{5.24}
$$

and

$$
\sum_{i=1}^{A} \frac{d_{i2}}{d_{i1}} > A
\tag{5.25}
$$

Otherwise, the area estimates will be A for one class and 0 for the other.

In addition, we state (without an extensive proof) that in the limiting case where $\sum_A \frac{d_{i1}}{d_{i2}}$ approaches A, the solution tend towards zero, such that:

$$
\sum_A \frac{d_{i1}}{d_{i2}} = A \Rightarrow p_1 = 0
\tag{5.26}
$$

and

$$
\sum_A \frac{d_{i2}}{d_{i1}} = A \Rightarrow p_2 = 0
\tag{5.27}
$$

Remembering our three small examples, we see that all three satisfy the first condition of the Lemma. $\mathbf{D}_1$ also satisfies the second, $\mathbf{D}_2$ does not and $\mathbf{D}_3$ is an example of the limiting case: $\frac{1}{4} + \frac{1}{2} + \frac{1}{2} + \frac{1}{1} + \frac{2}{1} + \frac{7}{4} = \frac{24}{4} = 6$. Finally, a special case occurs when

$$\sum_A \frac{d_{i1}}{d_{i2}} = A \text{ and } \sum_A \frac{d_{i2}}{d_{i1}} = A \tag{5.28}$$

This implies that all $e_i = \frac{d_{i1}}{d_{i2}}$ are equal to 1, because with $e_i = 1 + \delta$, such that:

$$
\begin{aligned}
\sum 1 + \delta_i &= \sum \frac{1}{1+\delta_i} \\[2mm]
\sum \frac{(1+\delta_i)^2}{1+\delta_i} &= \sum \frac{1}{1+\delta_i} \\[2mm]
\sum (1+\delta_i)^2 &= \sum 1 \\[2mm]
\sum (1 + 2\delta_i + \delta_i^2) &= A \\[2mm]
A + 0 + \sum \delta_i^2 &= A \\[2mm]
\delta_i &= 0
\end{aligned}
\tag{5.29}
$$

This means that the two probability densities are the same everywhere. The image contains no information upon which the two classes can be distinguished. $\Im_1(\mathbf{x}) \equiv 0$ for all p_1.

Why Does the Matrix D Satisfy the Conditions for a Non-Trivial Solution?

Suppose that the conditional probability density functions of our two classes are $F(\mathbf{x})$ and $G(\mathbf{x})$. Let A_k be the (unknown) number of pixels that actually belong to w_k, so that we have $A_1 + A_2 = A$. We want to knows the sum S_k of some characteristic $K(\mathbf{x})$ over the entire image, so that $S_k = \sum_A K$. If, for whatever reason, we prefer to work in the feature space, we must take the frequencies of occurrence $n_\mathbf{x}$ of each $\mathbf{x}$ into account, so

$$S_K = \sum_A K(\mathbf{x}) = \sum_X n_\mathbf{x} K(\mathbf{x}) \tag{5.30}$$

where X is the set of all possible feature vectors. Let $n_{1\mathbf{x}}$ and $n_{2\mathbf{x}}$ be the number of pixels with feature vector $\mathbf{x}$ in w_1 and w_2 respectively. They follow from the probability density functions

$$n_x = n_{1\mathbf{x}} + n_{2\mathbf{x}} = A_1 F(\mathbf{x}) + A_2 G(\mathbf{x}) \tag{5.31}$$

and therefore,

$$S_k = \sum_X A_1 F(\mathbf{x}) K(\mathbf{x}) + A_2 G(\mathbf{x}) K(\mathbf{x}) \tag{5.32}$$

Actually, we were looking for the sum of $K(x) = F(x)/G(x)$ and this now becomes

$$\sum_A \frac{F(x)}{G(x)} = \sum_X \left(A_1 F(x) \frac{F(x)}{G(x)} + A_2 G(x) \frac{F(x)}{G(x)} \right)$$

$$= A_1 \sum_X \frac{F^2(x)}{G(x)} + A_2 \sum_X F(x) \qquad (5.33)$$

$$= A_1 \sum_X \frac{F^2(x)}{G(x)} + A_2$$

Introducing $D(x)$ as $D = G - F$ and observing that $\sum_X D = \sum_X G - \sum_X F = 0$, we obtain

$$\sum_A \frac{F(x}{G(x} = A_1 \sum_X \left(\frac{(G+D)^2}{G} \right) + A_2$$

$$= A_1 \sum_X \left(\frac{G^2 + 2GD + D^2}{G} \right) + A_2$$

$$= A_1 \left(\sum_X G + 2 \sum_X D + \sum_X \frac{D^2}{G} \right) + A_2 \qquad (5.34)$$

$$= A_1 \left(1 + 0 + \sum_X \frac{D^2}{G} \right) + A_2$$

$$> A_1 + A_2 = A$$

We can prove the same for $\sum_A \frac{G}{F}$, and the two conditions in the Lemma in the previous section are satisfied. In the limiting case of, for example, $A_1 = 0$, this reduces to $\sum_A \frac{G}{F} = 0$. Both sums are 0 in the special case of $F = G$.

From 2 to N Classes

In dealing with N classes, instead of two, the same theory applies. To check the criteria for non-trivial solutions, according to our Lemma we take one class at a time, call it w_1, and group the other $N - 1$ classes into w_2 by averaging their probability densities. Remember, no assumptions were made about the shape of the probability density functions. As a result, we will get a total of $2N$ conditions and, again using one class at a time, we can show them to be satisfied according to the second part our theory. Note that the 'limiting case' of priors being equal to zero will not just be theoretical anymore. In the case of stratification, it will be more likely that only a subset of classes will occur in individual strata.

CHAPTER 6

Image Segmentation for Change Detection in Urban Environments

Hans-Peter Bähr

6.1 Introduction

There is an increasing demand by society for information on the urban environment. One of the most important challenges in this respect is to move from a 2.5D representation of urban areas to 3D and 4D — that is, including time. This can only be achieved using remotely-sensed images as the primary data source, in conjunction with the analytical techniques offered by Geographical Information Systems (GIS) and knowledge-based approaches. This Chapter presents a snapshot of work under way in this general area, focusing on three separate research projects being carried out at the author's institute, namely (i) change detection in the urban environment, (ii) map updating and (iii) multi-temporal modelling. The studies range in spatial scale. At the small scale, image data from Landsat Thematic Mapper (TM) and the ERS-1 Synthetic aperture Radar (SAR) are merged to help identify urban areas and to distinguish different categories of land cover/land use within them. The second study, which also makes use of data from Landsat TM, employs Delaunay triangulation to delineate the urban-rural boundary. This work is based on an initial land cover/land use map generated using a conventional multispectral classification algorithm applied to the TM data. Finally, at the large scale, we report current research to derive vector maps from digitized aerial photography. This is achieved using a Blackboard System, together with an associative memory or a Semantic Network (ERNEST), to extract and interpret the necessary spatial features, and taking into consideration 3D and generalization effects.

Broadly defined, remote sensing is not limited to digital, satellite-based sensor systems, but encompasses the realm of photogrammetry and hence conventional analogue sensors, including aerial photography. Viewed from this perspective, remote sensing has been used routinely to provide information on urban areas for analysis and planning purposes since the 1920s and 1930s. For instance, analogue ortho-photographs have long been used for urban planning. The question therefore arises: what can modern, digital remote sensing offer, over and above these traditional analogue techniques, to meet the requirements of present-day urban planners and of society in general. To answer this question, we must break it down into two parts, namely (i) the needs of planners and society, and (ii) the potential of technology in general, and remote sensing in particular, to meet these requirements.

As far as the needs of planners and society are concerned, it is important to recognize that these differ between developing countries and industrialized nations. They will be conditioned by local pressures which may arise as a result of a diverse set of factors, such as population growth or the limited space available for urban expansion. The remote sensing technology — both hardware and software — that we have at our disposal to address these requirements is, of course, developing continuously. Unfortunately, these technological developments have not always been motivated by the need to meet the specific needs of urban planners. This has resulted in an applications gap between their requirements, on one the hand, and technical capabilities, on the other. The gap is particularly evident for developing countries, where 'high-tech' fixes are not necessarily the most appropriate or, indeed, even a viable solution to the problems at hand. It is therefore an important challenge for science to provide the necessary tools, and the obligation for national governments is to arrange the necessary investment to facilitate their development.

In reality, the solution to these broad problems is found in a number of smaller, individual advances. In this context, one of the most pressing tasks for remote sensing is to make the step from analogue to digital data-processing. More specifically, attention needs to be given to:

a)   the digitization of existing data sets pertaining to the urban environment,

b)   the extension of analytical techniques from a consideration of urban areas in 2D to their representation in 2.5D and 3D,

c)   the ability to monitor changes in the urban environment (i.e. a further step from 3D to 4D), and

d)   the introduction of knowledge-based systems to assist in the decision-making process in relation to urban planning.

Each of these tasks presents very considerable challenges whose solution requires novel approaches to data collection and processing based on timely, consistent and objective spatial data sets. In this sense, the images recorded by Earth-orbiting and airborne remote sensing systems, particularly where they have stereoscopic observing capabilities, represent a relatively cheap tool for providing a detailed representation of the form and, to a limited extent, the functioning, of urban areas. In view of the very large volumes of data that these systems collect, however, it is essential that appropriate automatic and semi-automatic procedures for image restitution and segmentation are developed in order that maximum value can be derived from the data that they produce. This Chapter presents an overview of several research projects under way at the author's institution that seek to address this requirement.

Each is concerned with automatic image segmentation to provide information on urban areas appropriate to the planning community.

6.2 Small-Scale Data Acquisition

Although many of the important patterns and processes associated with urban areas operate at the large-scale[1], there is still a requirement for small-scale data acquisition, where the primary objectives is to *detect* urban areas; that is, to distinguish towns and cities from other spatial entities in the observed scene. A secondary objective may be to identify different types of land use within the urban areas. The capability of existing satellite-sensor systems to distinguish more than a handful of such land-use categories is currently relatively limited (see Barnsley *et al.*, this volume). Even so, given the cost and logistical problems associated with collecting such data by conventional means over large geographical regions, remote sensing remains a valuable tool. As such, it is important to develop techniques that can be used to extract the maximum information from the available data. In this context, we present two methods for image segmentation: the first of which takes a standard multispectral classification algorithm, but applies it to a combination of optical and microwave image data to improve the delineation of urban areas and their constituent land use categories; the second takes the output from this classification process and uses Delaunay triangulation networks to enhance the delineation of the urban-rural boundary.

6.2.1 The Multi-Sensor Concept for Satellite Image Classification

Description of the Test Area and Data Set

The city of Karlsruhe (275,000 inhabitants) and its hinterland in the Upper Rhine Valley were selected for this particular study. This flat, heavily populated urban area and its suburbs are surrounded by a number of isolated towns and villages. The total area considered is approximately $33km$ by $37km$ in size, which corresponds to 1325 by 1590 pixels in a $25m$ resolution raster data set. Table 6.1 shows the range of image data sets available to this study. These include five images recorded by the Landsat-TM sensor (in 1991 and 1993) and data from the Synthetic Aperture Radar (SAR) instruments on board the Japanese JERS-1 and European ERS-1 satellites. Note that the JERS-1 SAR operates in the C-band, while the ERS-1 SAR operates in the L-band. The data from these two sensors also differ in terms of polarization mode (data from the JERS-1 SAR were recorded in VV mode, while those from the ERS-1 SAR were acquired in HH mode), incidence angle (23.0° versus 35.2°, respectively) and date of acquisition (April, May and August). Consequently, the three SAR scenes were used as different phenomenological channels in the subsequent image classification process.

---

[1]The term *scale* is often used in different, sometimes contradictory ways, in the literature. Here, we use the term in the strict mapping sense. Thus 'large-scale' refers to a detailed or fine spatial resolution representation of surface features, rather than a large spatial coverage (i.e. large area) which, in mapping terms, would normally imply a 'small-scale' (i.e. comparative coarse spatial resolution) study.

Table 6.1: Data available for the study area.

Sensor	ERS-1	JERS-1	TM 93	TM 91
Dates	04.05.93	18.04.94 01.08.93	27.04.93 30.06.93 01.08.93	11.07.91 20.08.91
Spatial Resolution	25m	25m	30m	30m
Geocoded	yes	yes	yes	yes
Reception Rate	44	35	16	16
Channels	L-Band	C-Band	6	6
Polarization	HH	VV	-	-
Inclination Angle	97.7°	98.5°	98.2°	98.2°
Incidence Angle	35.2°	23.0°	-	-

Classification Stage

Eighteen candidate classes were used in the multi-sensor image classification, including flowing and standing water bodies (two classes), densely-populated and sparsely-populated residential areas (two classes), industrial districts, coniferous, deciduous and mixed forests (three classes), cereal crops (two classes), corn, sunflowers, root crops, soft fruit crops, grassland and unclassified. Initial experiments suggested that the radar imagery did not aid in the separation of the vegetation classes; consequently, a mask was applied to the data such that the vegetation classes were classified using the optical data only, while the remainder (including the urban categories) were classified using a combination of optical and microwave data. The results are presented in Figure 6.1.

Visually at least, the result of the multi-sensor classification appears to be very good, certainly better in our experience than using either the optical data or the microwave data alone. The city of Karlsruhe can be clearly seen in the lower right-hand corner of Figure 6.1. Moreover, the other towns and villages in this region are generally well represented, with far fewer errors of commission from the urban classes into the non-urban classes than is usually the case when the classification is based solely on optical data. The use of the microwave data also enables standing water (i.e. lakes, reservoirs and ponds) and flowing water (i.e. the Rhine) to be distinguished.

A further qualitative indication of the impact of the microwave data on the image classification process is given in Figure 6.2. The three frames in this figure focus on a small sub-scene extracted from the upper right-hand corner of the image in Figure 6.1, centered on the village of Graben-Neudorf. The left-hand and central frames present the results of image classification based solely on optical data (TM 1991 and 1993, respectively). The results presented in these images overestimate the extent of the built-up zone in this area. On the other hand, the results obtained using a combination of optical and microwave data (right-hand frame) provide a more accurate representation of the urban area. Finally, although they are not presented here, it is worth noting that land cover/land use classifications based on the microwave data alone yield relatively poor results.

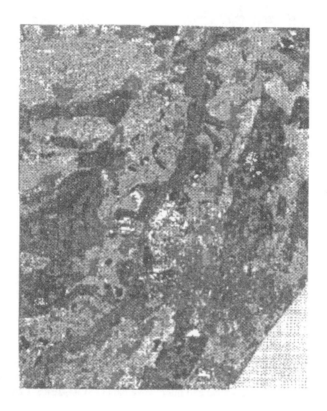

Figure 6.1: Results of conventional image classification applied to a combination of Landsat-TM (30/06/93, 6 channels), JERS-1 and ERS-1 SAR images. The data cover an area approximately $20km$ by $30km$ in size.

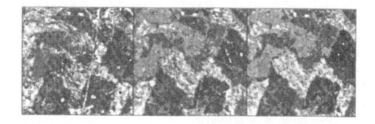

Figure 6.2: Land cover/land use classification of a small urban area and its non-urban periphery derived using optical image data alone (left and centre; based on 1991 and 1993 Landsat TM images, respectively) and in combination with microwave image data (right; based on 1993 Landsat TM and three SAR images).

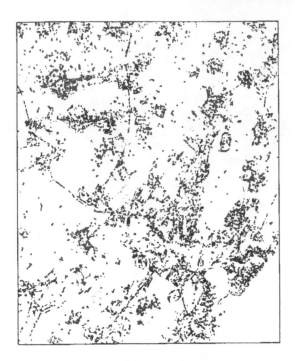

Figure 6.3: Binary image showing pixels which differ in terms of their assigned class (urban categories only) between two classifications of the Karlsruhe area: one based on optical data alone (TM 93), the other using optical and microwave data (TM 93, JERS-1 and ERS-1).

Quality Check

Rigorous evaluation of the output from image classification algorithms is vital. The optimum approach makes use of extensive field data for absolute verification. In many cases, however, this is not possible. In such circumstances, an alternative, but somewhat less powerful approach is to compare the results obtained from different classifications of the same scene. Figure 6.3 exemplifies this latter approach, through a comparison of the results obtained using optical image data (TM 1993) alone and a combination of optical and microwave data. It highlights the pixels which differ in terms of their assigned class (focusing on the urban classes only) between the two classifications.

There are several notable features evident in Figure 6.3. First, many linear elements are visible. This is partly a result of the greater ability of the microwave (c.f. optical) image data to detect man-made surface, such as roads and buildings, but it may also be a consequence of residual mis-registration between the optical and microwave images. Second, detailed analysis of the differences shown in Figure 6.3 reveals that many of the more extensive areas of difference are due to errors in the classification based on the optical data alone, which tends to over-estimate the extent of the urban areas. A much small number of the difference pixels are due to error in the classification of the combined optical and microwave image

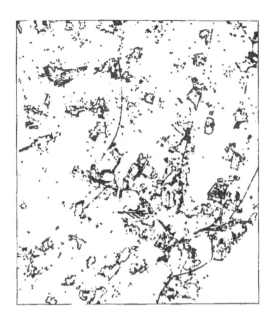

Figure 6.4: Differences (shown in black) between the urban land cover/land use classes extracted from a classification of optical (TM 93) and microwave image data, and the corresponding classes from the DSM.

data. Finally, it is of course possible that some pixels have been incorrectly classified in both the optical and optical plus microwave data sets. This type of error will not be evident in Figure 6.3, although it is known to exist, most notably in an area to the south of Karlsruhe.

A slightly more rigorous form of quality checking can be achieved by comparing the results of image classification with existing digital map data. In this study, we have used a 'digital situation model' (DSM) derived from 1:50,000 scale (TK 50) topographic maps, produced in 1985, for this purpose. The land cover/land use categories featured on this map include three urban classes, notably 'industry', 'transport' and 'settlements'. Analysis of the differences between the map and image data is complicated by the difference between the dates on which they were surveyed and acquired, respectively. Since the digital image data were acquired between six and eight years after the map was produced, we can expect real changes in land cover/land use to be highlighted, as well as errors in the image classification (and, potentially, in the digital map data). Bearing this point in mind, Figure 6.4 presents the differences between the urban land cover/land use classification based on the combined optical and microwave image data and the urban features presented in the DSM. The most notable aspect of this image is that many of the areas of apparent change occur around the boundaries of the urban areas reported in the DSM. This may be an artefact of the mixed pixels that occur at the border between two land cover/land use classes in the image data. Of perhaps greater significance, however, the solid patches of change suggest that certain areas have undergone urban expansion. One example of this, which has been verified in the field, occurs around the village of Friedrichstal (close to the right-hand margin of Figure 6.4, about half-way down the image). The southern extension of this village is clearly evident in the satellite data.

6.2.2   Delimitation of Urban Areas using Delaunay Triangulation to Update ATKIS

It is evident from the preceding discussion that the process of image classification is a complex one. In general, the best results are obtained where different, but complementary data sets are combined. This, however, demands accurate geometric rectification of the constituent images. Moreover, few of the data processing stages involved in image classification are fully (or even semi-) automated. The selection of sample areas for the training set, in particular, is a time-consuming manual task. Taken together, these form bottle-necks in the data-processing chain which restrict the widespread, operational application of remote sensing as a tool for providing information on urban areas. Full or partial automation of these tasks is therefore required to remove the bottle-necks. In this section, we explore the automation of one such task: the identification of the urban-rural boundary and, hence, the delimitation of urban areas. The approach that we outline moves beyond the pixel-oriented methods described earlier in this chapter, to include a consideration of *a priori* knowledge and neighbourhood relations. The approach also makes use of data from the ATKIS digital map system (Harbeck 1994), which is the official topographic mapping system in Germany.

Concept of the Approach

Urban areas typically exhibit different characteristics from all other classes in remotely-sensed images. These differences are partly expressed in terms of their spectral signatures on a pixel-by-pixel basis, but more importantly in terms the spatial variation, or heterogeneity, of this signal (see also Barnsley *et al.*, this volume). Conventional multispectral classification algorithms which assign each pixel to one of a set of candidate classes solely on the basis of its individual spectral response are unable to capture this heterogeneity. This has two implications: first, the per-pixel classification may contain errors as a result of the spatial variations in spectral response within urban areas (sometimes expressed as a 'salt-and-pepper' effect in the classified image); and second, these algorithms cannot exploit the information contained within the spatial heterogeneity to infer additional information about the urban environment, such as that pertaining to land use. To overcome this limitation, the method employed here combines a per-pixel examination of spectral response with semantic representations of the urban area based on neighbourhood relations/configurations.

Method

The spectral/semantic data-processing scheme used in this study is shown in Figure 6.5. The ATKIS digital map database is used to select training areas automatically for input to a conventional multispectral classification algorithm. To avoid the inclusion of pixels relating to vegetated areas (i.e. intra-urban open space) within the urban training sets, a simple vegetation index is computed for the image as a whole and used to remove pixels above a given threshold value. The resultant classification is used to identify all pixels

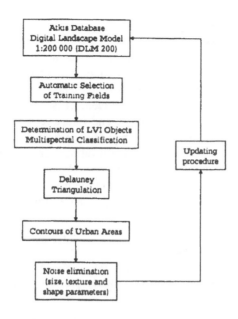

Figure 6.5: Method used to derive information on the extent of urban areas based on analysis of satellite sensor images, digital map data from the ATKIS database, and the use of Delaunay triangulation.

assigned to one of the urban categories. In general, we anticipate that urban areas will be characterized by tight spatial clusters of relatively large numbers of these pixels. By the same token, isolated 'urban' pixels that lie some distance from these cluster are probably not real urban areas, but are the result of error in the classification process. Such pixels can therefore be excluded from the final urban class. Various methods could be used to implement this procedure, including simple distance-decay models. Here, we have chosen to employ a Delaunay triangulation network to define the limits of the urban areas (Weindorf 1994). Urban areas are thus treated as a set of triangular spatial primitives. These contain semantic and topological information about the urban areas. For example, they can be used to eliminate error or 'noise' from the initial classification by taking into account *a priori* knowledge about the expected size and shape of the urban areas. Finally, the urban-rural boundaries derived using this method can be returned to the original ATKIS database for map updating.

Results

Figure 6.6 shows the results obtained using the data-processing sequence outlined above. Figure 6.6 displays the typical output of a per-pixel, multispectral classification algorithm. The black pixels in this image are candidate members of the urban area (i.e. they have been assigned to one of the urban spectral classes). The complex spatial pattern of these pixels is evident, as is the lack of a clearly defined urban-rural boundary. Visually, we can identify the locations of possible towns and village in this image on the basis of the dense spatial clustering of the black pixels. The automated identification of such clusters is performed using Delaunay triangulation (Figure 6.6b). An urban-rural boundary can be formed from an analysis of the Delaunay triangulation (Figure 6.6c). This is quite noisy, however, with disjoint roads and small, spurious urban areas included. These 'noise' features can be removed through a further analysis of the size, shape and image texture of the candidate regions (Figure 6.6d). Finally, in Figure 6.7 the output of the whole data-processing chain (Figure 6.6d) is compared to the corresponding urban-rural boundary stored in the ATKIS digital map data base. It is important to understand that the whole process up to the transition from Figure 6.6c to Figure 6.6d is fully automatic. On the basis of this preliminary experiment, the method appears to be very robust (resistant to noise) and identifies accurately the size and shape of urban areas.

The method described above has been applied to a more extensive study area (Figure 6.8) covering the city of Karlsruhe and its environs. The results appear very promising and, on the strength of this, the basic method is now being adapted to delineate different types of urban land use, such as industrial areas and densely-populated residential districts.

6.3   Large-Scale Data Acquisition from Aerial Photography

The distinction between 'small scale' and 'large scale' mapping employed in this Chapter is essentially one between satellite (small scale) and aerial (large scale) image data. Clearly, the higher spatial resolution generally associated with aerial imagery allows one to acquire more detailed information on urban areas pertaining, for example, to the size, shape and number of houses, roads, gardens etc.. However, the methods that must be used to derive this information are typically still more complicated than those used to analyze small-scale images. In this context, the use of digital map data to support and inform the analysis of aerial images is very common.

There are, however, important differences between images and maps (Quint and Bähr 1994). Figure 6.9 shows an example comparing map and image data covering a small section of Karlsruhe University campus. The white lines derived from the map data (original scale 1:5,000) do not match exactly with the boundaries of the corresponding objects in the image (original scale 1:6,000). The differences are due to the radial displacement of 3D objects away from the principal point, and the effects of residual error in the geometric registration of the map and image. Other reasons include gross errors in the map data and changes in the surface features since the map was compiled. Despite these differences, a trained human photo-interpreter can readily match both representations. Achieving the same thing using computer software is a more challenging task. In general, it requires us to transform both the map and the image data into a higher symbolic level. Two different approaches to this problem are described in the next two sections.

6.3.1   Blackboard Approach

The first approach is known as the 'blackboard' method and is based on an idea first presented by Newell (1962). Figure 6.10 shows the overall architecture employed in this ap-

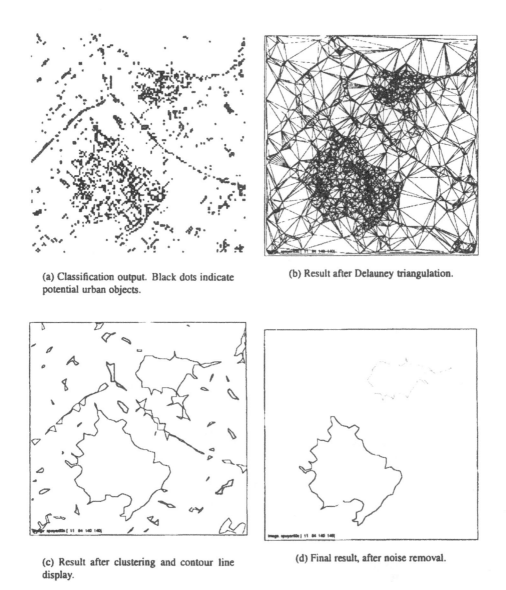

(a) Classification output. Black dots indicate potential urban objects.

(b) Result after Delauney triangulation.

(c) Result after clustering and contour line display.

(d) Final result, after noise removal.

Figure 6.6: Results of urban-rural boundary extraction using the Delaunay triangulation method.

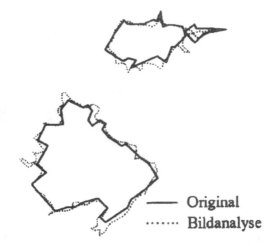

Figure 6.7: Comparison of the final results from the Delaunay triangulation method and the urban-rural boundary stored in the ATKIS database. Solid line = ATKIS; dotted line = results of image analysis.

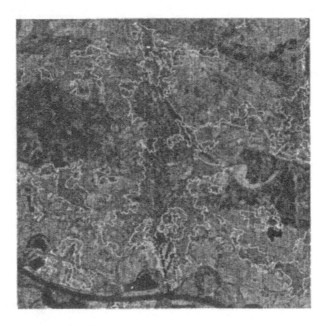

Figure 6.8: Urban contour lines for the City of Karlsruhe and its environs.

Figure 6.9: Differences between an aerial image and the representation of the same spatial features in digital map data.

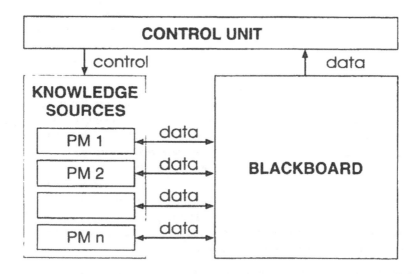

Figure 6.10: Blackboard architecture.

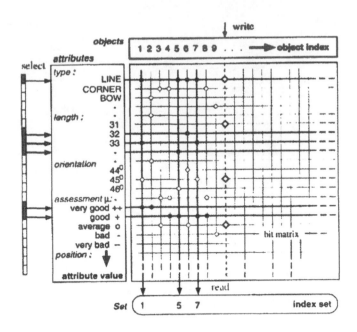

Figure 6.11: Associative memory structure.

proach, including the associative memory or 'blackboard'. A control unit puts data onto the blackboard from a set of 'knowledge sources', as well as reading information from the blackboard. Different objects in a map or image are put into the columns of the associative memory (Figure 6.11), which stores attributes of those objects such as their edges, corners, arcs, links, orientation, quality etc.. This approach is very effective in terms of both read and write procedures.

Two other, important aspects of the blackboard approach are the matching of scene primitives and the construction of more complex objects from simpler ones. Figure 6.12 gives an example of the former. It relates to a search-procedure to identify the four corners of rectangular objects. The many potential corners detected in an image or a map are put into the blackboard together with their attributes (e.g. orientation). A systematic search of the set of corners which might form a rectangle yields n^k test samples, where n is the number of elements in the scene and k the number of elements required to form the object (four in the case of corners of a rectangle). For the aerial photograph shown in Figure 6.9 this yields 79,727,040,000 samples! This number can be reduced considerably by eliminating subsets of corners which clearly cannot form a rectangle.

In practice, the modelling of urban objects can also be done in much more sophisticated ways, including the extension to 3D. Figure 6.13, for example, shows how a building might be models out of rectangles in a 3D scene. The geometric conditions may be pre-determined on the base of mathematical conditions.

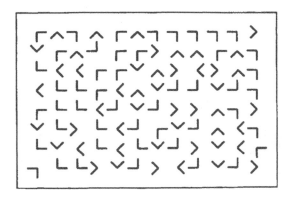

Figure 6.12: Composition of rectangular objects (e.g. houses) from graphic primitives in 2D (after Stilla *et al.* 1995).

6.3.2 Semantic Network Approach

The second procedure used to segment maps and images at the author's institute involves semantic networks. Here, the information content of the observed scene is represented in the form of a graph (see also Barnsley *et al.*, this volume). Individual objects form nodes of the graph, sometimes referred to as 'concepts'. The relations between the concepts represented by the edges of the graph. Figure 6.14 shows an example semantic network for part of an urban scene, produced using the ERNEST software developed at the author's institute (Niemann *et al.* 1990). As with the blackboard approach, semantic networks can be used to model specific objects that form part of the urban area. Figure 6.15, for example, shows the model — in the form of a graph — of the concept 'garden'. Both the blackboard system and semantic network yield a symbolic description of the scene.

6.4 Conclusions

Urban areas are by their very nature complex. Although a human operator can extract information from images of urban areas relatively easily, computer-based automated interpretation is a challenging task. Even apparently simple problems, such as determining the extent of the urban area as a whole, require data from multiple sources (including, but not restricted to, different types of satellite sensor). In general, we note that at larger mapping scales, more complex and sophisticated modelling tools are required to reconstruct the spatial location, 3D geometry and type of urban objects.

6.5 References

Harbeck, R., 1994, Das Geoinformationssystem ATKIS und seine Nutzung in Wirtschaft und Verwaltung, Technical report, Bonn-Bad Godesberg, Landesvermessungsamt Nordhein-Westfalen.

110

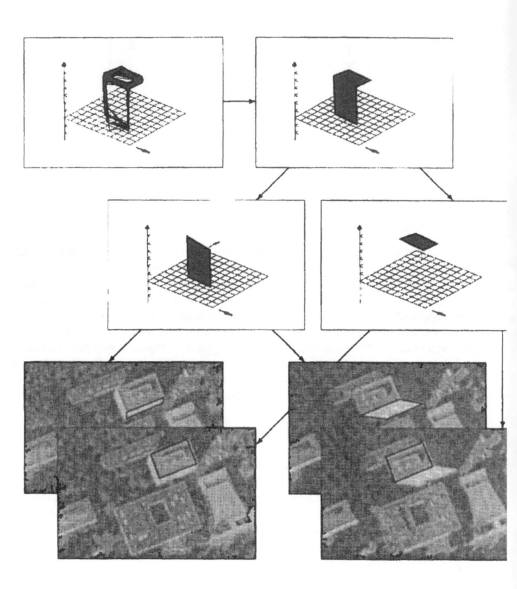

Figure 6.13: Extraction of building objects in 3D (after Stilla *et al.* 1995).

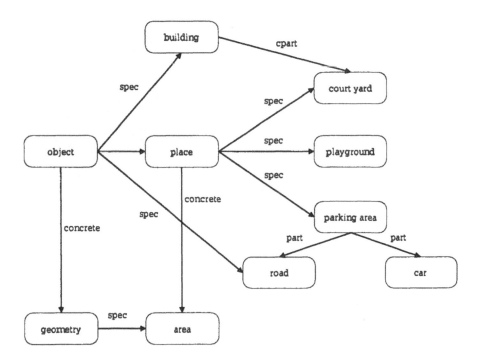

Figure 6.14: An example semantic network (after Quint and Bähr 1994).

112

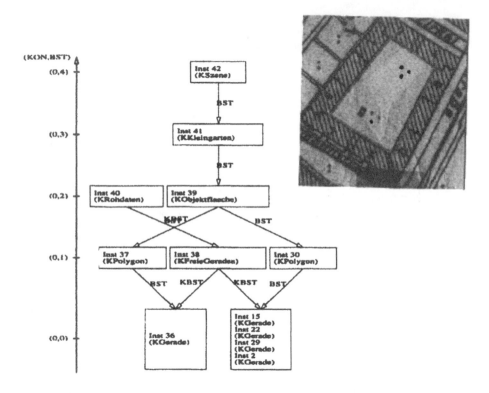

Figure 6.15: Description of a 'garden' object using the ERNEST semantic network software.

Newell, A., 1962, Some problems of basic organization in problem-solving programs, In *Proceedings of the Second Conference on Self-Organizing Systems*, edited by M. C. Yovits (Spartan Books).

Niemann, H., Sagerer, G., and Schröder, S., 1990, ERNEST: A semantic network system for pattern understanding. *IEEE Transactions on Pattern Analysis and Machine Intelligence.*

Quint, F., and Bähr, H.-P., 1994, Feature extraction for map-based image interpretation, Technical report, Laboratory for Information Engineering in Surveying, Mapping and Remote Sensing, Wuhan, China.

Weindorf, M., 1994, *Segmentation von Siedlungsflächen aus Satellitenbilddaten*, Master's thesis, Universität Karlsruhe, Karlsruhe, Germany.

CHAPTER 7

Inferring Urban Land Use by Spatial and Structural Pattern Recognition

Michael J. Barnsley, Lasse Møller-Jensen and Stuart L. Barr

7.1   Introduction

Approximately 85% of the European Community's population lives and works in urban areas (Eurostat 1993). This spatial concentration of human activity has very significant environmental and economic impacts, both within and beyond the immediate urban fabric, arising from the need to service it with appropriate physical resources (e.g. energy, food and water) and to dispose of the resultant waste products. Despite this, basic information on urban areas — such as their location, physical extent, human population and rate of growth — is often dated, inaccurate or simply non-existent. In this context, the potential for satellite remote sensing systems to deliver timely, consistent and spatially comprehensive data sets seems clear. In practice, however, there has been a marked disparity between the apparent promise and delivered performance (Barnsley *et al.* 1989, Kutsch-Lojenga and Meuldijk 1993, Orsi 1993). To ensure that greater value is derived from satellite remote sensing in the future, it is instructive to examine the causes of this disparity.

Early satellite sensors, notably the Multispectral Scanning System (MSS) on board the Landsat series of satellites, were quite poorly adapted to provide anything other than the most general information about urban areas (Jackson *et al.* 1980, Forster 1980). In part, this was a function of the limited number of broad spectral wavebands in which Landsat MSS recorded data. More importantly, the relatively coarse spatial resolution (*c.*80*m*) of this sensor was generally inadequate to provide an accurate delineation of the urban-rural boundary or to identify and distinguish land cover/land use types in urban areas (Barnsley and Barr 1996). This led many observers at the time to highlight sensor spatial resolution as the principal limiting factor for urban studies (Forster 1985, Toll 1985). By implication, results would improve when finer spatial resolution data became available.

With the launch of the Thematic Mapper (TM) sensor on board Landsat-4 in 1984 and the High Resolution Visible (HRV) instruments on board SPOT-1 in 1986, the civilian remote sensing sector was provided access to multispectral images with a spatial resolution of $30m$ and $20m$[1], respectively. Paradoxically though, many of the initial studies that made use of these data reported *reduced* levels of accuracy from conventional (i.e. per-pixel) multispectral classifications of urban areas (Forster 1985, Toll 1985, Haack *et al.* 1987, Martin *et al.* 1988). The explanations for this were typically expressed in terms of 'scene noise' (Gastellu-Etchegorry 1990) — that is, many areas which had appeared to be spectrally homogeneous in data recorded by earlier satellite sensors were now revealed as being spectrally heterogeneous in images from the new, finer spatial resolution devices — the implication being that the problem lay in the inherent spatial complexity of urban scenes and the improved ability of the new satellite sensors to resolve their component elements (e.g. buildings, roads and various types of open space). This explanation is, however, at best unhelpful, at worst misleading. The spatial variation in detected spectral response to which it refers is controlled, in part at least, by the size, shape and spatial arrangement of buildings, roads and different types of intra-urban open space. Far from being unwanted 'noise', the resultant pixel-to-pixel variability is in fact the signal — a potential source of information on these and other properties of the urban scene. The 'problem' of scene noise therefore derives more from the inability of standard image processing techniques (e.g. per-pixel multispectral classification algorithms) to extract useful information from the data, than from the intrinsic limitations of the data themselves (Barnsley and Barr 1996).

If existing data-processing strategies are deficient, alternatives must be sought. This problem has become even more pressing in recent years, with the launch of a new generation of optical satellite sensors capable of acquiring images at still higher spatial resolutions. Examples include the LISS-III sensor ($\sim 6m$ in panchromatic mode) on board the current IRS-1C satellite and the imaging instruments ($1m - 4m$) on board the IKONOS-1 satellite (Aplin *et al.* 1997, Ridley *et al.* 1997). Data from these devices clearly offer exciting new opportunities for monitoring urban areas from space. At the same time, they create very considerable challenges in terms of developing data-processing techniques to exploit the information that they contain. In this context, we note that the spatial variability inherent in urban scenes, typically expressed as *texture* in images acquired by Landsat-TM and SPOT-HRV, will be more clearly revealed as *pattern* in digital data from the new, very fine spatial resolution sensors.

7.2   From Land Cover to Land Use

Remote sensing specialists are frequently rather lax in their use of the terms 'land cover' and 'land use', often applying them interchangeably or mixing examples of each within a single classification scheme. Their meanings are, however, distinct. Broadly speaking, *land cover* refers to the physical materials on the surface of a given parcel of land (e.g.

---

[1]The SPOT-HRV sensors also provide data with a spatial resolution of $10m$ in their panchromatic mode.

grass, concrete, tarmac, water), while *land use* refers to the human activity that takes place on, or makes use of, that land (e.g. residential, commercial, industrial). The fundamental problem for remote sensing is that while there is often a relatively simple, direct relationship between land cover type and detected spectral reflectance, the same is seldom true for land use: land use is an abstract concept, an amalgam of cultural and economic factors, most of which cannot be determined directly by means of remote sensing. This problem is central to studies of urban areas, since land use is normally the property of greater interest. Having said that, many categories of urban land use have a characteristic spatial pattern of spectrally-distinct land cover types that enables their recognition in fine spatial resolution remotely-sensed images. For example, residential districts in many Western European towns and cities typically comprise a complex assemblage of buildings (houses), roads and open space (gardens and parks). Human photo-interpreters, of course, use information on the size, shape, relative proportions and spatial arrangement of these and other scene elements as 'cues' to identify different types of urban land use. What is required, then, is some means of formalizing this process and of embedding it within an automated or semi-automated, digital image-processing system.

It has been suggested that one way in which this might be achieved is to divide the image analysis process into two distinct stages (Barnsley and Barr 1996): the first involving a low-level segmentation of the image into a set of discrete, labelled regions, each delineating an area of homogeneous land cover; the second encompassing some procedure to infer the principal land use associated with groups of one or more these regions on the basis of characteristic spatial textures, patterns or assemblages of land cover (Gurney and Townshend 1983, Møller-Jensen 1990, Gong and Howarth 1992a, Gong and Howarth 1992b, Eyton 1993, Barnsley and Barr 1996). Barr and Barnsley (1997) have expressed this more formally as the composition of two mappings, $f : I \mapsto F$ and $g : F \mapsto S$,

$$I \circ S = \{(i \mapsto s) : i \in I, s \in S, \exists f \in F([i \mapsto f] \wedge [f \mapsto s])\} \tag{7.1}$$

where

I is the set of multispectral responses for each of the pixels in the image,

F is the set of land cover classes (or other first-order themes), and

S is the set of land use categories (or other second-order themes).

In principle, any one of a large number of techniques can be used to perform the first of these two mappings, ranging from standard (i.e. per-pixel) multispectral classification algorithms (both supervised and unsupervised) to artificial neural networks. The relative merits of, and problems associated with, each these techniques is discussed extensively elsewhere (see, for example, Mesev *et al.*, Chapter 5 this volume) and will not be rehearsed further here. Instead, it is important to note two general points:

- The accuracy with which the multispectral responses are mapped onto the appropriate land-cover classes will affect the success with which land use information can be inferred from these data — in other words, error in the first mapping will be propagated through to the second.

- Many of the techniques that are appropriate to the second mapping require that the regions are labelled in terms of meaningful land cover types — this generally militates against the use of unsupervised classification algorithms in the first stage.

Several different methods have been developed to infer land use from an analysis of the spatial, textural and contextual arrangement of land cover types present within an image, three of which are considered in detail in this Chapter, namely

1. empirical/statistical kernel-based techniques,

2. knowledge-based texture models, and

3. region-based structural pattern recognition techniques.

7.3   Case Study 1: Statistical, Kernel-Based Techniques

One of the simplest ways to characterize the spatial patterns of land cover in a classified image, with the aim of relating these to different categories of urban land use, is to pass a convolution kernel (or moving-window filter) across the data. Several kernel-based techniques have been developed that are suitable candidates for this purpose. Wharton (1982), for example, used the frequency of class labels within an $m \times n$ pixel kernel to infer the dominant urban land use associated with the central pixel in the window. This approach approach has since been applied to fine spatial resolution images from Landsat-TM and SPOT-HRV (Gong and Howarth 1992a, Gong and Howarth 1992b, Eyton 1993).

In this section, we examine the results obtained from a modified version of Wharton's basic method, developed by Barnsley and Barr (1996), in which the spatial arrangement of the land cover labels within the $m \times n$ pixel window is analyzed, in addition to their relative frequencies. Barnsley and Barr (1996) argue that this allows more subtle differences in urban land use to be distinguished; for example, between different *densities* of residential land. Their method, referred to as SPARK (*SPA*tial *Re*-classification *K*ernel), determines the matrix, M_{ij}, of adjacency events for land cover classes i and j between every pair of adjacent pixels within the current window (Figure 7.1). The value of each element of this matrix denotes the frequency with which pixels belonging to class i are adjacent to those belonging to class j within the kernel. The matrix M_{ij} is then compared to 'template' matrices, T_{kij}, determined from training areas of known land use categories, k, sampled within the image. Equation 7.2 gives the index used to assess the degree of similarity between the spatial pattern and frequency of land cover labels for the current position of the kernel and the candidate land-use categories:

$$\Delta_k = 1 - \sqrt{\frac{1}{2N^2} \sum_{i=1}^{C} \sum_{i=j}^{C} (M_{ij} - T_{kij})^2} \tag{7.2}$$

where

M_{ij} is an element of the current adjacency-event matrix,

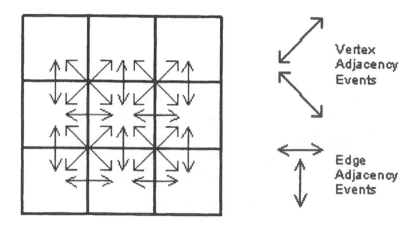

Figure 7.1: Diagrammatic representation of the set of adjacency events within a 3 × 3 pixel kernel (after Barnsley and Barr 1996).

T_{kij} is the corresponding element of the template matrix for land use category k,

N is the total number of adjacency events within the window ($N = 20$ for a 3 × 3 kernel), and

C is the number of land-cover classes in the image.

Finally, the central pixel in the window is assigned the label of the land use category for which Δ_k is maximized (Barnsley and Barr 1996).

The SPARK algorithm has been applied to a SPOT-HRV multispectral image centred on the town of Swanley, Kent, UK (Barnsley *et al.* 1995). This work formed part of a larger project entitled "Remote Sensing and Urban Statistics", organized by Eurostat (the Statistical Office of the European Communities), to examine the potential of satellite remote sensing to provide detailed statistical information on urban land use (Barnsley *et al.* 1995, Eurostat 1995, Eurostat 1998). An important aspect of this project was that the criteria used were established by Eurostat, in conjunction with the National Statistical Institutes (NSI) of the participating member states[2], rather than by remote sensing specialists. This included the use of a specially developed land-use nomenclature, known as CLUSTERS (*C*lassification for *L*and *U*se *ST*atistics *E*urostat's *R*emote *S*ensing project; 7.1). Although attention was given to the types of information that might be derived from satellite sensor images, the CLUSTERS scheme was primarily designed with the needs of European planners and statisticians in mind. It therefore provides a stringent test of the capabilities of both existing and newly-developed image processing techniques.

---

[2]The participating members states and NSIs in the first phase of the project were the U.K. (Department of the Environment, Transport and the Regions), France (INSEE), Germany (Statistisches Bundesamt) and the Netherlands (Central Bureau voor de Statistiek).

Table 7.1: Extract from the CLUSTERS land use nomenclature scheme.

Level I		Level II		Level III		Level IV	
A	Man-made areas	A1	Residential areas and public services	A11	Residential areas	A111	Continuous and dense residential areas
						A112	Continuous residential areas of moderate density
						A113	Discontinuous residential areas of moderate density
						A114	Isolated residential areas
						A115	Collective residential areas
				A12	Public services, local authorities	A120	Public services, local authorities
		A2	Industrial or commercial activities	A20	Industrial or commercial activities	A201	Heavy industry
						A202	Manufacturing industrial activities
						A203	Commercial and financial activities and services
						A204	Agricultural holdings
		A3	Technical and transport infrastructures	A31	Technical infrastructures	A311	Technical networks, protective structures
						A312	Waste and water treatment
				A32	Transport	A321	Road transport, rail networks
						A322	Airports and aerodromes
						A323	River and maritime transport
		A4	Extractive industries, building sites, tips and wasteland	A41	Extractive industry	A410	Extractive industries
				A42	Building sites, tips and wasteland	A421	Building sites
						A422	Tips
						A423	Wasteland

Summary results from the application of the SPARK algorithm and their subsequent integration with other spatial data sets within a GIS are presented in Figure 7.2. This shows a comparison of the extent of the 'built-up' area determined from the remotely-sensed images (shaded area) and from contemporaneous digital map data held by the UK Department of the Environment, Transport and the Regions (DETR)[3]. The methods used to process the satellite images are described in detail by Barnsley *et al.* (1995), but briefly they involve the application of the SPARK algorithm followed by a series of automated procedures to remove spurious outliers and transportation links (e.g. roads and railways) that are not included within the Eurostat definition of urban areas. The degree of correspondence between the satellite-derived 'built-up' areas and the conventionally surveyed, digital map data is generally quite good (91%; Barnsley *et al.* 1995), although there are clearly areas in which the satellite-derived product underestimates the true extent of the urban area (usually in places where the density of buildings is very low) and others where it overestimates it (generally due to mis-classification of fallow fields surrounding the urban area and the inclusion of various transportation network features that are not formally part of the 'built up' zone).

The primary advantages of SPARK are its conceptual simplicity and ease of implementation. Unfortunately, it suffers from the deficiencies of all kernel-based techniques, namely (i) it tends to smooth the boundaries between discrete land cover/land use parcels, (ii) it is difficult to determine *a priori* the optimum size for the kernel, and (iii) a regular, rectangular window represents an artificial area within which to search for the spatial pattern of irregularly-shaped land cover/land use parcels (Dilworth *et al.* 1994, Fisher 1995, Barnsley and Barr 1996). Barnsley and Barr (1996) suggest a number of potential enhancements that might be made to the basic SPARK algorithm to overcome the first two problems identified above, including:

- the use of an adaptive window-size for the kernel to take account of the different spatial scales of variation in land cover characteristic of different land use categories, and

- the use of information on the probability/confidence of class-label assignment output on a pixel-by-pixel basis by maximum likelihood classification algorithms during the first mapping stage (see, for example, Mesev *et al.*, Chapter 5 this volume).

The third problem, however, remains. Moreover, the 'model' used to represent the spatial composition of land cover types within each land use category is determined empirically and must therefore be recomputed for different scenes and for images of the same scene acquired by satellite sensors with different spatial resolutions. Thus, we believe that the scope for further development of kernel-based techniques for urban land use mapping is limited. Their applicability to the new sources of very fine spatial resolution ($< 5m$) image data is also questionable. For example, because of the very small area covered by each pixel in such images, a very large kernel would be required to capture the spatial pattern of land cover representative of any given land use. Unfortunately, the use of large kernel-sizes inevitably results in blurring and smoothing of the boundaries between land cover/land use parcels, which is generally undesirable in most urban mapping and monitoring applications.

---

[3] Known at the time of the study at the Department of the Environment (DoE).

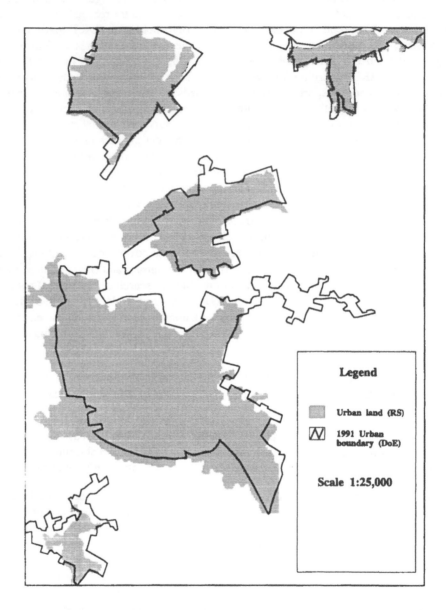

Figure 7.2: 'Built-up' areas (shaded) of Swanley and surrounding towns in Kent, UK derived from a 1992 SPOT-HRV multispectral image using the SPARK algorithm. The figure also shows the corresponding, conventionally surveyed, digital map data of the urban-rural boundary.

7.4   Case Study 2: Knowledge-Based Texture Models

The second case study that we examine here takes a rather different approach to inferring urban land use. It recognizes that, even with the latest generation of satellite sensors, it is not possible to resolve completely and unambiguously all of the land cover objects (e.g. buildings, roads and different types of open space) present in urban areas. Specifically, the precise extent and shape of these objects, and by implication the boundaries between them, will be uncertain owing to spectral mixing within the image pixels (i.e. 'mixed pixels'). One possible solution to this problem would be to try to unmix the spectral signatures of the component scene elements and hence to reconstruct more precisely their relative areal coverages, if not their original shape and boundaries (Quarmby *et al.* 1992, Ichoku and Karnieli 1996, Foody *et al.* 1997). The alternative approach adopted in this study is to relate the resulting spatial variations in detected spectral response (i.e. image texture) to the dominant land use. The fundamental assumption underlying this approach is that different types of urban land use will be expressed as different textures in the corresponding satellite-sensor image depending on, for example, the size, shape and density of their constituent buildings, roads and areas of open space. Rather than simply attempting to correlate image texture and urban land use using a statistical model, however, the approach employed here is based on formal spatial models of urban structure and knowledge-based techniques.

7.4.1   Spatial Object Models of Urban Scenes

Interpretation of remotely-sensed images, whether by humans or computers, is arguably most effective when it is based on a sound knowledge of the object, and possibly the context, of analysis. In the domain of urban areas, this implies knowledge of the constituent elements of the urban scene (e.g. the size and shape of buildings, roads and open space) and an understanding of the ways in which their spatial composition differs between various categories of urban land use. This domain-specific knowledge can, in principle, be represented through simple spatial models of urban structure. Used in conjunction with databases on the spectral and spatial properties of the constituent scene elements, the purpose of such models is to help delineate areas of different urban land use by predicting their appearance in remotely-sensed images.

The spatial object model referred to above may be either generic or specific in nature. A generic model attempts to capture the features common to an entire class of objects, usually at the expense of their individual details. Such a model reflects what might be described as the 'class typical' appearance of the object — a pure form which may not necessarily be found within a particular scene. A specific model, on the other hand, attempts to represent the detailed characteristics of one particular object.

Based on a generic model, the algorithm used to classify the scene into discrete land use parcels must be able to measure the degree of similarity between the observed region and the spatial object model. There is, of course, a trade-off between generality and specificity. The advantages of a generic model are that (i) it may be possible to construct a relatively simple model to represent the properties of a large number of objects of analysis, and (ii) the model may be portable from one location to another with only minor modifications. On the other hand, since the match between the object model and the patterns observed in the image

Table 7.2: Attributes stored for each object class in the model.

Object name
Parent object name
Object code (national coding system is used)
Object size / certainty factor
Object colour / certainty factor
Object shape (length/width)

will generally not be exact, there is typically greater uncertainty associated with the land use label assigned to each region than for a more specific model.

7.4.2   Implementation of the Spatial Object Model

In this case study we describe the use of a prototype spatial object modelling system implemented in the Prolog programming language. Prolog is particularly well suited to the kind of structured knowledge representation outlined above, notably through the use of 'frames' or semantic networks (Hendrix 1979, Ringland and Duce 1989). Thus, the spatial model may be implemented as a number of nodes and connecting links, in which each node represents an object class and the links represent the various relations held between the classes (Figure 7.3).

The prototype system used in this study allows certain *properties* of a specific object of analysis to be described (Table 7.2). The most important of these are the estimated size, colour and shape of the object, and the estimated number of instances of each object class. These attributes have been selected to describe some of the basic physical properties that are relevant in the context of remote sensing. The spatial model contains several types of object class *relation* that convey information on spatial context, the most important being the *APO* relation (meaning '*a part of*' or '*spatially included in*'). Knowing that one object class is a part of another is, however, insufficient to model satisfactorily the spatial properties of a real urban scene. This is because it provides no information about the number of subordinate objects (subparts) within the object class or their spatial distribution within the scene. To overcome this problem, and hence make the model more generic, an estimate — even a very approximate one — of the expected number of objects in a sub-class can be provided by the user and stored as an attribute of the sub-class. The spatial model can be made more specific by providing heuristic knowledge about the spatial location of two object classes relative to one another. Thus, it is possible to specify *adjacent to* and *apart from* relations between two objects. The former can be extended to take into account the percentage of the total perimeter of the two classes which is common to both of them. Inheritance of properties from objects at another level is also possible under certain conditions.

To limit data redundancy and to facilitate updating, the system also allows other types of objects and relations to be defined. These can act as a form of common database. This is useful when a number of object classes share the same attributes (e.g. identical spectral properties), in which case it is inefficient to store the same information several times in the model. Thus, for example, the *has surface* (*HS*) relation links all urban object classes

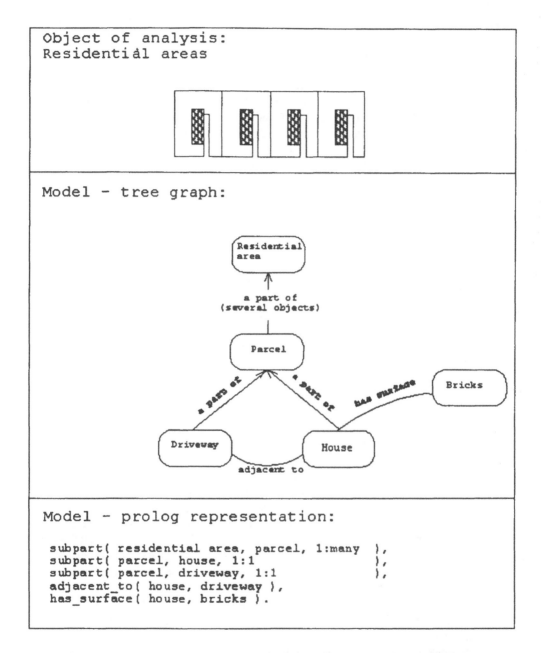

Figure 7.3: Representation of a residential urban area: object of analysis (top), corresponding spatial model visualized as a tree graph showing the available link types (middle), and its Prolog representation (bottom).

with similar surface types to a single database that contains information about the spectral properties of that surface[4]. A number of other databases can be added to the system. For example, the prototype system includes a sensor database which enables it to access, among other things, information about the spatial, spectral and temporal resolution of a potential image data source.

Finally, the prototype system also stores certainty (uncertainty) information about the object attributes (Table 7.2). This has been introduced for two reasons. First, the attribute value may be an estimate (or guess) on behalf of the user: perhaps because the real value is not known or is too difficult to measure. Secondly, the attribute value may represent the average for a class of objects which has a large within-class variation. This may result in a substantial difference between the model value of the attribute and the actual value for the majority of objects in that class. The second type of uncertainty is part of the general problem of inferring from a generic spatial object model, outlined above.

7.4.3 Modelling Image Texture

Using the information contained within the spatial object model, it is possible to simulate the textural appearance of an urban land use class in a remotely-sensed image of a given spatial resolution and spectral waveband. In other words, the textural manifestation of a composite urban (land use) object can be predicted from knowledge of (i) its component sub-parts (i.e. land cover parcels), (ii) their spectral properties, and (iii) their size, shape and spatial arrangement within the scene. It is then possible to search for regions of similar textural appearance in the remotely-sensed image in an attempt to identify areas of the candidate land use category.

To do this, the spatial object model must be mapped onto some measure (or measures) of image texture. Here, we use a variety of measures based on standard, grey-level co-occurrence matrices (Haralick et al. 1973, Haralick 1979). These contain information about the frequency with which neighbouring pixels hold values A and B, respectively. This information is accumulated across all possible pixel-pairs within the region of interest. Neighbourhood relations are defined in all four directions relative to the target pixel (i.e. pixels touching along an edge, but not at a vertex). An A–B relation is considered identical to a B–A relation. The system allows each object class from the spatial object model to be entered into the co-occurrence matrix one at a time, starting at the most detailed level. Thus, the textural appearance of a complex, compound object class can be constructed within the co-occurrence matrix by accumulating the responses of its component sub-classes. This requires that the system makes use of model information concerning object hierarchy, the expected number of objects, the spatial distribution of objects, etc. Once the co-occurrence matrix is constructed for a given object class, a number of second-order texture measures (e.g. contrast, entropy and angular second moment) can be computed (Haralick et al. 1973, Haralick 1979).

Although this describes the general way in which object classes are entered into the co-occurrence matrix, it is instructive to examine this process in somewhat more detail. For

---

[4]In the prototype system employed here, these relate to the mean and standard deviation of the pixel values for that surface type.

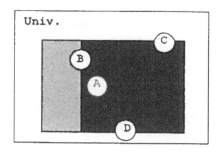 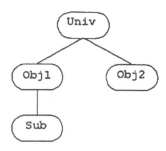

Figure 7.4: Four types of border relations between objects. A: obj1–obj1, B: obj1–obj2, C: obj1–univ and D: sub–univ.

each object, the system computes the expected number of internal pixel-pairs (i.e. where the pixels represent the same object) and border pixel-pairs (i.e. where the pixels represent different objects). Currently, the system is only able to handle rectangular objects. The expected length and width of the object, expressed in pixels, are computed from the absolute object size and the shape index attribute, while the number of internal and border pixel-pairs are computed as follows:

$$\text{internal pixel pairs} = W(L - 1) + L(W - 1) \tag{7.3}$$

$$\text{border pixel pairs} = 2L + 2W \tag{7.4}$$

where L is the expected length of the object, and W is its expected width[5]. Finally, it is assumed that the variation in textural properties that might be expected for different instances of objects in a generic class can be approximated by a bivariate Gaussian distribution of values within the corresponding co-occurrence matrix.

The spatial autocorrelation of pixel values in the 'internal' areas of an object is normally very high, at least for the target objects examined in this study, and this should be reflected in the co-occurrence matrix. The number of internal pixel-pairs 'occupied' by each sub-class is therefore subtracted from the total number of internal pixel-pairs of a given object class. The result is entered directly into the matrix using an estimated mean, standard deviation and autocorrelation of the pixel values for that class. It should be noted that the prototype system does not handle the task of converting surface reflectance properties into estimated pixel values. Instead, the pixel values employed in the test runs described in the results section are extracted from the actual digital image and entered directly into the model using the HS relation.

---

[5]The methods used to handle objects smaller than a single pixel will be described later in this section.

Four types of border relation may exist between objects (Figure 7.4). As the generic model does not contain precise information about the nature of these border relations, however, they have to be estimated by the system. The *apart from* and *adjacent to* relations supplied by the user are employed, if present; otherwise the default relation is adjacent to, provided that the total size of all sub-classes is greater than 75% of the size of their super-class, and *apart from*, if not. Once all border relations of a sub-class are established the expected number of pixel-pairs covering two object classes is computed. To account for mixed pixels, it is estimated that 25% of each border pixel is influenced by the other pixel in the pair and the pixel values are adjusted accordingly.

Handling Mixed Pixels

Object classes that are sub-pixel in size clearly cannot be entered directly into the co-occurrence matrix, since the matrix is generated from values associated with complete pixel-pairs. It is therefore necessary to distinguish between object classes depending on whether the sizes of the current class and its super-class (defined by the *part of* relation) fall above or below some critical threshold. The threshold value is set to one pixel for the current class and four pixels for its super-class. This implies that it is impossible to identify reliably and measure the texture of objects smaller than these thresholds. In these circumstances, the system behaviour is modified as follows:

- if the super-class is smaller than four pixels in size, the model is altered so that the spectral value of the super-class includes the contribution from the current class, which is subsequently deleted from the model;

- if the current class is smaller than one pixel, but its super-class is larger than four pixels, a new object class is generated by the system. Each object of the new class (NC) is given the size of one pixel. The spectral value of that pixel is determined by the influence of the current class on one pixel of the super-class. The current class is then deleted from the model and the NC is evaluated and entered into the co-occurrence matrix.

7.4.4   Results and Discussion

The prototype system described above has been used to test the extraction of spatial information on urban land use categories from remotely-sensed images acquired at different spatial resolutions. Two test areas located in Accra, Ghana, each dominated by a single land use, were selected for this purpose. Area 1, the Cantonments, is a low-density residential district dominated by large land-cover parcels, extensive vegetation, isolated building units and a clear road pattern. Area 2, the Nima-Mamobi region, is a very high-density residential district with compound houses and a few trees. A black-and-white aerial photograph of each area was scanned at a pixel resolution of $0.7m \times 0.7m$. These data were then resampled to $10m \times 10m$ and $30m \times 30m$ to simulate images acquired by the SPOT-HRV (in panchromatic mode) and Landsat-TM satellite sensors, respectively. This provides a good test of the ability of the prototype system to handle data acquired at three very different levels of spatial resolution, using essentially the same object model. It also encompasses the range of

Table 7.3: Second-order texture values generated from the raw image and from the model for $0.7m \times 0.7m$, $10m \times 10m$ and $30m \times 30m$ pixels.

0.7m Resolution	Area 1		Area 2	
Texture:	Image	Model	Image	Model
Contrast	34.83	34.82	62.12	68.03
Angular Second Moment	0.58	0.58	0.29	0.31
Entropy	6.74	6.65	7.14	7.12
10m Resolution	Area 1		Area 2	
Texture:	Image	Model	Image	Model
Contrast	375.30	307.45	181.07	183.15
Angular Second Moment	0.19	0.15	0.35	0.24
Entropy	7.64	7.97	6.95	7.31
30m Resolution	Area 1		Area 2	
Texture:	Image	Model	Image	Model
Contrast	316.08	360.56	61.39	126.35
Angular Second Moment	0.34	0.34	1.13	0.50
Entropy	6.87	7.17	5.70	6.51

resolutions typical of the older generation of satellite sensors, in which the basic urban objects are normally smaller than the size of the image pixels, and those of the new generation of satellite sensors, in which the urban objects are typically larger than the image pixels.

The spatial model was established by estimating the object sizes and spectral features from the original aerial photograph. The choice of object class hierarchy is, of course, rather subjective and may be expressed in several different ways. The process of estimating spectral features directly from the data source is not entirely in agreement with the general objectives of the system, which calls for knowledge-based description of the object of analysis, but it serves the purpose of demonstrating the functionality of the system. The approach for testing the texture modelling is to compare co-occurrence matrices based on generic spatial object models with corresponding matrices computed from the image data. This is done for three standard, second-order texture features, namely contrast, entropy and angular second moment(Haralick 1979). The results are presented in Table 7.3.

Before analyzing these results, it is important to consider a number of general points. First, to have confidence in the ideas underlying the general approach outlined above, it is important that the modelled and computed texture values pertaining to the same object (area) should be similar: this implies that the model provides a faithful representation of the actual scene structure. At the same time, model values should differ significantly for objects (areas) with different spatial structures. Secondly, it is implicitly assumed that sufficient pixels exist in the image data for a meaningful textural pattern to emerge: this may not be the case in the lowest resolution data set, such that the derived texture values may be more erratic and less predictable.

The results presented in Table 7.3 suggest that there is indeed a close correspondence between the modelled texture values and those derived from the image data of each test area, while the modelled values differ substantially between the two test areas. The exception to this general rule is for Area 2 (the Nima-Mamobi region) in the 30m spatial resolution data, where the measured and modelled texture values differ by 100%. This is probably a result of the limited number of pixels available at this spatial resolution, which prevents a reliable texture value from being calculated. For the same reason, the correspondence between the measured and modelled values becomes stronger as the spatial resolution of the data increases.

The results obtained from the prototype system are therefore very positive and offer considerable encouragement for its further development and operational application. Clearly, the prototype system involves a number of compromises and restrictions — it is a far from complete solution. Nevertheless, it provides an important test-bed for further research and offers a clear pointer to what might be achieved in terms of urban land use mapping using very high spatial resolution satellite sensor images.

7.5   Case Study 3: Structural Pattern-Recognition Techniques

The final approach to inferring urban land use from remotely-sensed images that we examine here makes use of region-based, structural pattern-recognition techniques. As will be seen, there are strong parallels between this approach and the knowledge-based texture models described in the previous section. Both conceive of the urban scene as comprising a set of interrelated spatial objects, and both attempt to represent the properties of, and relations between, those objects. They differ, however, in the sense that the knowledge-based texture models take an essentially top-down approach to image analysis (modelling the *expected* appearance of the urban scene and then searching for regions of the image that match this model), whereas the region-based, structural pattern-recognition techniques described in this section adopt a bottom-up approach (quantifying the structural composition of land cover parcels in the image and then comparing this to areas of known land use). Compared to per-pixel, statistical and neural pattern-recognition methods, the uptake of structural (or syntactic) approaches by the remote sensing community has been quite limited. The potential benefits of this approach are considerable, however, especially in the context of data acquired by the latest generation of very fine ($< 5m$) spatial resolution satellite sensors, and it seems likely that it will be a focus of greater research and development activity in the future.

The example of the region-based, structural pattern-recognition approach that will be examined here is provided by Barr and Barnsley (1997), who have developed a system known as SAMS (Structural Analysis and Mapping System). SAMS operates on ordinal or categorical raster-format data — typically a land-cover classification generated from a remotely-sensed image. It can be used to derive structural information about thematically-labelled regions (e.g. land cover parcels) present in such data. The boundaries of each region are identified using a simple contour-tracing algorithm (Gonzalez and Wintz 1987, Gonzalez and Woods 1993). These are represented using Freeman chain-codes (Freeman 1975) and are stored in a Region Search Map (RSM). The RSM can be processed to derive further information about the structural characteristics of the observed scene, including various morphological properties of the constituent regions (e.g. their area, perimeter and various measures of their shape), as well as the spatial and structural relations between them (e.g. adjacency, containment, distance and direction). This information is used to populate a graph-theoretic data model, known as XRAG (eXtended Relational Attribute Graph), which is defined by the heptuple:

$$XRAG = \{N, E, EP, I, L, G, C\} \tag{7.5}$$

where

N is the set of nodes (i.e. regions), such that $N \neq \oslash$;

E is the set of (extrinsic) spatial relations between $n \in N$ (e.g. *adjacency* and *containment*);

EP is the set of properties associated with the relations in E (e.g. *distance* and *direction*);

I is the set of (intrinsic) properties relating to $n \in N$ (e.g. *area* and *perimeter*);

L is the set of labels (interpretations) assigned to $n \in N$ (e.g. *grass*, *tree*, *urban*, *non-urban*, etc.);

G is the set of groups binding $\forall l \in L$ to the context of a scene interpretation (e.g. the label *tarmac* is 'bound' to the group *land cover*, while the label *residential* is bound to the group *land use*); and,

C is the set stating the confidence to which $l \in L \to n \in N$ (e.g. "the probability that region n belongs to land cover class Y is 0.9").

Each region in the RSM file is represented by a *node*, $n \in N$, in the XRAG data model. The existence of a relationship between a pair of regions n_x and n_y for a given relation, $r_i \in E$, is represented by an *edge* (i.e. $(n_x, n_y) \in r_i$). Thus, the node set N, in combination with the relation r_i, is equivalent to a standard relational graph, $\{N, r_i\} \equiv G$ (Barr and Barnsley 1997). Non-relational properties of the regions (e.g. their *area* and *perimeter*) are represented as attributes of the nodes, while relational properties (e.g. the *distance* and cardinal *direction* (*orientation*) between any two regions) are represented as attributes of the edges.

    Since the preceding discussion is rather abstract, it may help to illustrate the basic functionality of SAMS/XRAG using a simple example. Suppose that we have produced a raster-format land-cover map of an imaginary urban area which, for the purpose of this Chapter, we shall call '*Graphtown*' (Figure 7.5). Let us also assume that the land cover map for *Graphtown* has been produced from a fine spatial resolution remotely-sensed image and that the resultant classification is entirely free from thematic error. SAMS identifies the twelve discrete land-cover regions within the scene (Figure 7.6) and stores the spatial location and exterior boundaries of these regions within the Region Search Map (RSM). A number of morphological properties of these regions, as well as the spatial and structural relations between them, can be determined from the RSM and used to populate the XRAG data model.

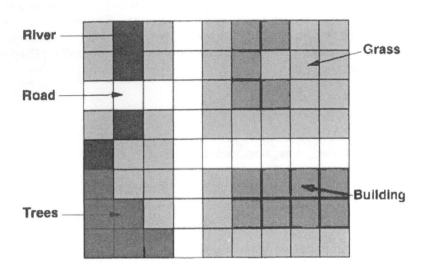

Figure 7.5: Land cover map of *Graphtown*.

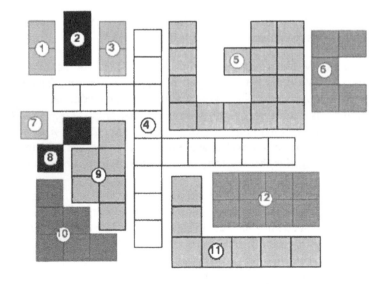

Figure 7.6: 'Exploded' representation of *Graphtown* showing the twelve discrete land-cover regions identified within the scene.

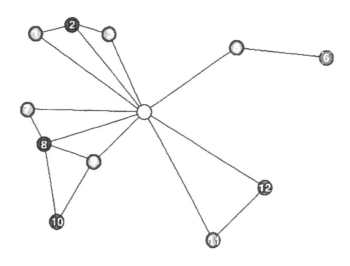

Figure 7.7: Graph visualization of *Graphtown* for the spatial relation *adjacency*.

Figure 7.7, for example, shows a graph visualization of *Graphtown* for the spatial relation *adjacency*. This shows, for example, that region 8 is adjacent to regions 4, 7, 9 and 10.

Initial experiments with SAMS/XRAG concentrated on the identification of the urban-rural boundary in SPOT-HRV multispectral images (Barr 1992, Barr and Barnsley 1995), but more recently the focus has been on the potential for information extraction from finer spatial resolution data sets (Barnsley and Barr 1997). To this end, a number of data-processing algorithms has been developed which can be used to interrogate the XRAG data model, to process the information contained within it, and to update the contents of the data model accordingly. Preliminary tests have also been performed on SAMS/XRAG by Barnsley and Barr (1997) using land cover parcels generated from Ordnance Survey 1:1,250-scale digital map data (Land-Line.93+). Selected features (i.e. roads, buildings, woodland, water bodies and open space) were extracted from these data and used to produce a simple thematic map. This vector coverage was then converted into raster format, generating a land cover 'image' with an effective spatial resolution of 1m (Figure 7.8). Figure 7.9 shows the adjacency graph produced from these data. It should be evident from this diagram that, while the overall spatial structure for this scene is remarkably complex, there are distinct 'clusters' of nodes and edges, each of which exhibits a somewhat different structural pattern. In exploring this point further, Barnsley and Barr (1997) identified a number of test areas within the main scene, each representative of a particular category of urban land use (Figure 7.10). They showed that simple, quantitative measures could be derived from the information stored in the XRAG data model that would allow certain categories of urban land use to be distinguished on the basis of differences in their structural composition (Figure 7.11) and the morphological properties of their component land-cover parcels (Barnsley and Barr 1997).

Figure 7.8: Ordnance Survey 1:1,250-scale Land-Line.93+ digital map data covering part of the town of Orpington in the Borough of Bromley, south-east London, U.K. The data have been topologically structured, labelled and converted to a 1m raster.

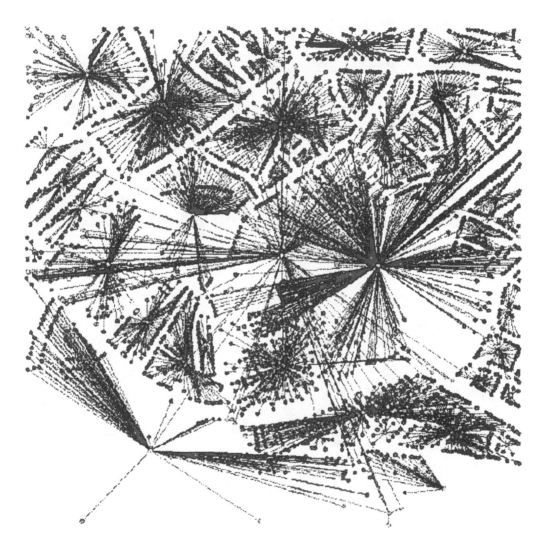

Figure 7.9: Graph visualization of Figure 7.8 for the spatial relation *adjacency*.

136

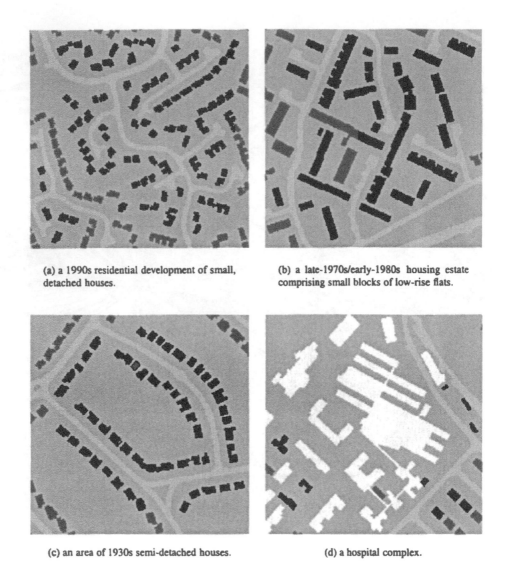

(a) a 1990s residential development of small, detached houses.

(b) a late-1970s/early-1980s housing estate comprising small blocks of low-rise flats.

(c) an area of 1930s semi-detached houses.

(d) a hospital complex.

Figure 7.10: Sample areas of four different types or age of land use selected from the raster-format thematic map presented in Figure 7.8.

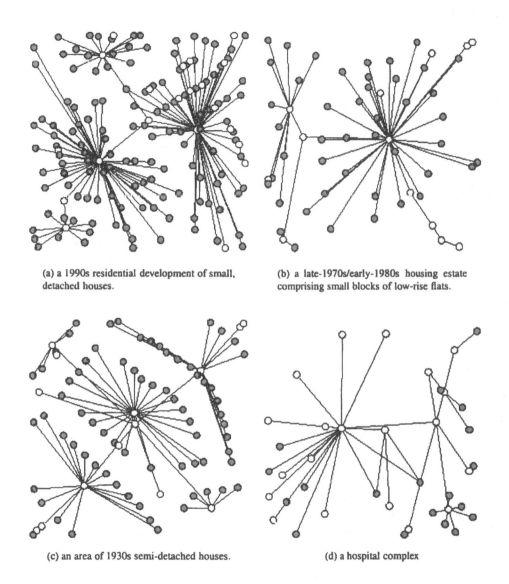

(a) a 1990s residential development of small, detached houses.

(b) a late-1970s/early-1980s housing estate comprising small blocks of low-rise flats.

(c) an area of 1930s semi-detached houses.

(d) a hospital complex

Figure 7.11: Graph visualization of the spatial relation *adjacency* for sample areas of four different types or age of land use presented in Figure 7.10.

Of course, tests performed on digital map data do not provide a particularly good indication of what might be achieved using fine spatial resolution remotely-sensed images, since the former are the product of various levels of feature selection and abstraction, as well as thematic and spatial generalization. In general, the spatial structure of remotely-sensed images will be much more complex because they are influenced by, among other things, differences in roofing and road-surface materials, as well as variations in vegetation type, cover and health. The resultant land cover data will also be subject to the effects of mixed pixels, shadowing, occlusion and misclassification (Barr and Barnsley 1999). These factors will affect the morphological properties of, and the spatial relations between, the derived land cover parcels. This will, in turn, complicate attempts to infer land use from the spatial composition of land cover.

To address this problem, we have begun to develop a number of reflexive-mapping techniques that can be used to remove — or, at least, to reduce — much of the structural 'clutter' in the initial land-cover maps produced from fine spatial resolution remotely-sensed images (Barr and Barnsley 1998). Figure 7.12 shows preliminary results obtained using these techniques, concentrating on the derived 'built' regions. A visual comparison of Figure 7.12a, which shows the 'built' objects extracted from the original Ordnance Survey digital map data, and Figure 7.12b, which shows the corresponding regions derived from a $2m$ spatial resolution remotely-sensed image of the same area, illustrates the extent of the structural clutter problem. Almost all of the 'built' regions present in the digital map data are evident in the remotely-sensed image, but there is also a very large number of typically small, spurious 'built' regions (i.e. 'clutter' regions). Figure 7.12c presents the results of the reflexive-mapping procedure, which suggest that it is capable of removing most of these clutter regions, while preserving the morphological properties and spatial pattern of the true 'built' regions. Finally, for the purpose of comparison, Figure 7.12d presents the results obtained by applying a conventional, post-classification majority filter to the original land cover classification. Although this has the effect of removing many of the 'clutter' regions, it fails to preserve the size and shape properties of the true 'built' regions and, hence, the spatial relations between them. A more rigorous, quantitative analysis of the relative performance of these techniques is currently underway.

Ultimately, the purpose of deriving and representing structural data is to infer additional information about the corresponding scene — for example, that pertaining to land use (Barnsley and Barr 1997). Relatively little attention has, however, been given to the development of the data-processing techniques needed to perform this type of inference based on images obtained by Earth-orbiting satellite sensors. Indeed, there is a general paucity of formal models on the structural operators, their semantics and expected results, required to underpin the development of such techniques. While a full discussion of the necessary models and techniques is beyond the scope of this Chapter, a number of general approaches that might act as a starting point for the development of structural inference tools can be highlighted.

Several techniques have been developed to assess the similarity between graph models and graphically-represented spatial data (Schalkoff 1992). In general, these examine whether the model and the data are *isomorphic* (i.e. have the same graph structure) or, if they are not, the extent to which they are non-isomorphic (e.g. by determining the number of structural differences that exist) (Ballard and Brown 1982, Schalkoff 1992, Sonka *et al.*

(a) 'built' regions extracted from the 1m spatial resolution land cover map covering part of the town of Orpington in the Borough of Bromley, south-east London, UK.

(b) 'built' regions extracted from a land cover classification of a 2m spatial resolution multispectral image of the same area.

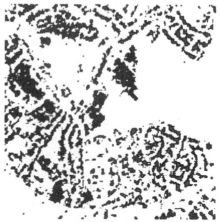

(c) 'built' regions extracted from (b) after applying the reflexive-mapping technique.

(d) 'built' regions extracted from (b) after applying a conventional, post-classification majority filter using a 3 × 3 pixel kernel.

Figure 7.12: Results of structural 'clutter' reduction using a reflexive-mapping technique within SAMS/XRAG and conventional post-classification majority filtering.

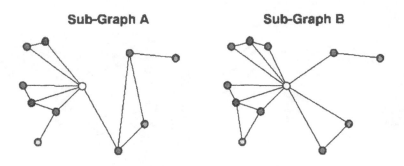

Figure 7.13: Comparison of sub-graphs.

1993, van der Heijden 1994, Figure 7.13). Graph matching and graph similarity measures have been widely used in a number of computer vision/machine vision tasks, though these often involve a relatively small number of basic structural descriptions of the spatial entities in the observed scene (Ballard and Brown 1982, Sonka *et al.* 1993, van der Heijden 1994). Their applicability to the analysis and interpretation of images acquired by Earth observation sensors is likely to be more problematic, largely because of the greater structural complexity of the corresponding scenes. This results in much greater variability — and, hence, uncertainty — not only in terms of the geometric properties of scene primitives (regions), but also in their structural properties and relations.

7.6   Conclusions

This chapter has focused on the production of land use data for urban areas from satellite sensor images. In particular, it has highlighted the very considerable potential of the new generation of commercial satellites which will have on board very fine spatial resolution optical sensors. These will offer an unprecedented opportunity for mapping and monitoring urban areas from space. It was suggested that developments in sensor technology must, however, be matched by equivalent advances in the techniques available to process and interpret the data that the new sensors produce. Of course, simple, visual interpretation remains a possibility, but novel automated or semi-automated techniques may be required if we are to derive the maximum potential benefit from the very large volumes of data that will soon be generated by these Earth-orbiting sensors. Three possible approaches to this problem have been explored in this chapter, the first makes use of kernel-based (i.e. 'moving window') techniques, the second employs knowledge-based approaches to examine image texture, while the third applies region-based, structural pattern-recognition methods to infer urban land use. Each has its advantages and disadvantages, and each is at a slightly different stage of development. It is our strong sense that the latter two approaches offer the greatest potential for the future, principally because they are founded on explicit models of the physical structure and composition of urban areas. Both draw heavily on techniques originally developed in the field of computer vision/machine vision, although they have been adapted to address the specific demands and constraints of Earth Observation data and urban scenes. Clearly, much remains to be done before either approach can routinely provide maps of urban land use from satellite sensor images, but the process of achieving this goal has at least begun.

7.7 Acknowledgements

The authors would like to acknowledge the support of U.K. Natural Environment Research Council through the provision of research grant number GR3/10186. The digital map data used in this study are reproduced from the Ordnance Survey 1:1,250-scale Land-Line.93+ series with the kind permission of the Controller of Her Majesty's Stationary Office, Crown Copyright.

7.8 References

Aplin, P., Atkinson, P. M., and Curran, P. J., 1997, Fine spatial resolution satellite sensors for the next decade. *International Journal of Remote Sensing*, **18**, 3873–3882.

Ballard, D. H., and Brown, C. M., 1982, *Computer Vision* (New York: Prentice Hall).

Barnsley, M. J., and Barr, S. L., 1996, Inferring urban land use from satellite sensor images using kernel-based spatial reclassification. *Photogrammetric Engineering and Remote Sensing*, **62**, 949–958.

Barnsley, M. J., and Barr, S. L., 1997, A graph-based structural pattern recognition system to infer land-use from fine spatial resolution land-cover data. *Computers, Environment and Urban Systems*, **21**, 209–225.

Barnsley, M. J., Barr, S. L., and Sadler, G. J., 1995, Mapping the urban morphological zone using satellite remote sensing and GIS: UK involvement in the Eurostat Pilot Project on remote sensing and urban statistics, In *Proceedings of the 21st Annual Conference of the Remote Sensing Society* (Nottingham: Remote Sensing Society), pp. 209–216.

Barnsley, M. J., Sadler, G. J., and Shepherd, J. W., 1989, Integrating remotely sensed images and digital map data in the context of urban planning, In *Proceedings of the 15th Annual Conference of the Remote Sensing Society*, Remote Sensing Society (Nottingham: Remote Sensing Society), pp. 25–32.

Barr, S. L., 1992, Object based re-classification of high resolution digital imagery for urban land use monitoring, In *International Archives of Photogrammetry and Remote Sensing, Commission VII* (Washington, D.C.: ISPRS), pp. 969–967.

Barr, S. L., and Barnsley, M. J., 1995, A spatial modelling system to process, analyse and interpret multi-class thematic maps derived from satellite sensor images, In *Innovations in GIS 2*, edited by P. Fisher (London: Taylor and Francis), pp. 53–65.

Barr, S. L., and Barnsley, M. J., 1997, A region-based, graph-theoretic data model for the inference of second-order information from remotely-sensed images. *International Journal of Geographical Information Science*, 11, 555–576.

Barr, S. L., and Barnsley, M. J., 1998, Removing structural clutter from land-cover classifications of very high spatial resolution images using reflexive-mapping techniques, In *Proceedings of ECO BP'98, International Symposium on Resource and Environmental Monitoring, International Archives of Photogrammetry and Remote Sensing, Commision VII*, ISPRS (Budapest, Hungary: ISPRS), pp. 315–322.

Barr, S. L., and Barnsley, M. J., 1999, A syntactic pattern recognition paradigm for the derivation of second-order thematic information from remotely-sensed images, In *Advances in Remote Sensing and GIS Analysis*, edited by P. Atkinson, and N. Tate (Chichester: John Wiley and Sons), pp. 167–184.

Dilworth, M. E., Whister, J. L., and Merchant, J. W., 1994, Measuring landscape structure using geographic and geometric windows. *Photogrammetric Engineering and Remote Sensing*, 60, 1215–1224.

Eurostat, 1993, *The Impact of Remote Sensing on the European Statistical System — Proceedings of the Seminar, Bad Neuenahr* (Luxembourg: Office for Official Publications of the European Communities).

Eurostat, 1995, *Pilot Project Delimitation of Urban Agglomerations by Remote Sensing: Results and Conclusions* (Luxembourg: Office for Official Publications of the European Communities).

Eurostat, 1998, *The Impact of Remote Sensing on the European Statistical System — Proceedings of the Seminar, Esquilino, Rome* (Luxembourg: Office for Official Publications of the European Communities).

Eyton, J. R., 1993, Urban land use classification and modeling using cover-type frequencies. *Applied Geography*, 13, 111–121.

Fisher, P. (editor), 1995, *Innovations in GIS 2* (London: Taylor and Francis).

Foody, G. M., Lucas, R. M., Curran, P. J., and Honzak, M., 1997, Non-linear mixture modelling without end-members using an artificial neural network. *International Journal of Remote Sensing*, 18, 937–953.

Forster, B. C., 1980, Urban residential ground cover using Landsat digital data. *Photogrammetric Engineering and Remote Sensing*, 46, 547–558.

Forster, B. C., 1985, Principal and rotated component analysis of urban surface reflectances. *Photogrammetric Engineering and Remote Sensing*, 51, 475–477.

Freeman, J., 1975, The modeling of spatial relations. *Computer Graphics and Image Processing*, 4, 156–171.

Gastellu-Etchegorry, J. P., 1990, An assessment of SPOT XS and Landsat MSS data for digital classification of near-urban land cover. *International Journal of Remote Sensing*, **11**, 225–235.

Gong, P., and Howarth, P. J., 1992a, Frequency-based contextual classification and gray-level vector reduction for land-use identification. *Photogrammetric Engineering and Remote Sensing*, **58**, 423–437.

Gong, P., and Howarth, P. J., 1992b, Land-use classification of SPOT HRV data using a cover-frequency method. *International Journal of Remote Sensing*, **13**, 1459–1471.

Gonzalez, R. C., and Wintz, P., 1987, *Digital image processing* (New York: Addison-Wesley).

Gonzalez, R. C., and Woods, R. E., 1993, *Digital image processing* (New York: Addison-Wesley).

Gurney, C. M., and Townshend, J. R. G., 1983, The use of contextual information in the classification of remotely sensed data. *Photogrammetric Engineering and Remote Sensing*, **49**, 55–64.

Haack, B., Bryant, N., and Adams, S., 1987, An assessment of Landsat MSS and TM data for urban and near-urban land-cover digital classification. *Remote Sensing of Environment*, **21**, 201–213.

Haralick, R. M., 1979, Statistical and structural approaches to texture. *Proceedings of the IEEE*, **67**, 786–804.

Haralick, R. M., Shanmugan, K., and Dinstein, I., 1973, Textural features for image classification. *IEEE Transactions on Systems, Man and Cybernetics*, **8**, 610–621.

van der Heijden, F., 1994, *Image based measurement systems: Object recognition and parameter estimation* (Chichester: John Wiley).

Hendrix, G., 1979, Encoding knowledge in partitioned networks, In *Associative networks — the representation and use of knowledge in computers*, edited by N. Findler (New York: Associated Press).

Ichoku, C., and Karnieli, A., 1996, A review of mixture modelling techniques for sub-pixel land cover estimation. *Remote Sensing Reviews*, **13**, 161–186.

Jackson, M. J., Carter, P., Smith, T. F., and Gardner, W., 1980, Urban land mapping from remotely-sensed data. *Photogrammetric Engineering and Remote Sensing*, **46**, 1041–1050.

Kutsch-Lojenga, F., and Meuldijk, D., 1993, Developments towards spatial statisics in Europe, In *The Impact of Remote Sensing on the European Statistical System — Proceedings of the Seminar, Bad Neuenahr* (Luxembourg: Office for Official Publications of the European Communities), pp. 11–19.

Martin, L. R. G., Howarth, P. J., and Holder, G., 1988, Multispectral classification of land use at the urban-rural fringe using SPOT satellite data. *Canadian Journal of Remote Sensing*, **14**, 72–79.

Møller-Jensen, L., 1990, Knowledge-based classification of an urban area using texture and context information in Landsat-TM imagery. *Photogrammetric Engineering and Remote Sensing*, **56**, 899–904.

Orsi, A., 1993, Remarks based on meetings held in various countries (France, Greece, Italy, Portugal, Spain), In *The Impact of Remote Sensing on the European Statistical System — Proceedings of the Seminar, Bad Neuenahr* (Luxembourg: Office for Official Publications of the European Communities), pp. 21–29.

Quarmby, N. A., Townshend, J. R. G., Settle, J. J., White, K. H., Milnes, M., Hindl, T. L., and Silleos, N., 1992, Linear mixture modelling applied to AVHRR data for crop area estimation. *International Journal of Remote Sensing*, **13**, 415–425.

Ridley, H. M., Atkinson, P. M., Aplin, P., Muller, J.-P., and Dowman, I., 1997, Evaluating the potential of forthcoming commercial U.S. high-resolution satellite sensor imagery at the Ordnance Survey. *Photogrammetric Engineering and Remote Sensing*, **63**, 997–1005.

Ringland, G., and Duce, D. (editors), 1989, *Approaches to knowledge representation: an introduction* (Research Studies Press).

Schalkoff, R. J., 1992, *Pattern Recognition: Statistical, Structural and Neural Approaches* (New York: John Wiley).

Sonka, M., Hlavac, V., and Boyle, R., 1993, *Image Processing, Analysis and Machine Vision* (London: Chapman and Hall).

Toll, D. L., 1985, Effect of Landsat Thematic Mapper sensor parameters on land cover classification. *Remote Sensing of Environment*, **17**, 129–140.

Wharton, S. W., 1982, A context-based land-use classification algorithm for high- resolution remotely sensed data. *Journal of Applied Photographic Engineering*, **8**, 46–50.

CHAPTER 8

Urban Agglomeration Delimitation using Remote Sensing Data

Christiane Weber

"A city is a too complex thing to be defined without misunderstanding" (G.Perec)

8.1 Introduction

The third millennium will be urban. Today, almost 50% of the world's population is urban. Forty years of exceptional demographic change have had impacts upon the features of cities all over the world. This phenomenon has specific manifestations over the five continents in terms of population concentration, the density of buildings, economic activities and services, transportation development and traffic flows, etc. But what are the similarities between Atlanta and Calcutta, Florence and Bamako? Do cities have the same cognitive representation, the same reality? Do cities share similar historical heritages or growth trends?

In a few years, the locus of city development will shift from the industrial and other, old European countries to cities in Asia and the developing countries. Of course, the reasons for these developments are not uniform, yet this shift has profound implications for the international network of cities. Cities are no longer isolated in the middle of nowhere, but rather belong to one or more city networks, and are inextricably linked in evolutionary terms. To be a part of the network is crucial, in order to ensure social and economic development, and membership criteria are defined in terms of population size, economic activity rates, employment characteristics, attractiveness, etc. In order to assess any city's prospects, it is important to be able to compare cities. This is quite difficult to do because definitions and realities differ. Regarding the specifics of today's city networks, we might focus upon the following features:

- the advent of conurbations which gather together millions of inhabitants of different, but joined cities — such as the north-eastern seaboard of the US or the Japanese coast;

- the domination of regions by their major cities, which encourages commuting, con-
  centrates all activities, and starves the surrounding areas of urban functions — as in
  the case of the Paris region;

- the densification of the urban fringes of administrative urban units, to the detriment
  of agricultural hinterlands and natural landscapes; and

- a rural-urban continuum of urban units or agglomerations, where differences are based
  more on the dominant type of living space (through the extensive or intensive use of
  space) than on city morphology, social composition or demographic dynamics.

What kind of realities are hidden behind these words and concepts? What criteria should
be developed to provide acceptable definitions of these concepts for the majority of people,
knowing that 'non rural' is the only obvious blanket categorization? The definition of a
'city' or 'agglomeration'[1] invites two general considerations (and associated criteria): first,
a specified population and delimited space and, secondly, economic and social relationships
with contiguous surroundings. These considerations are difficult enough to classify when
dealing with a national settlement system, but they become still more difficult when dealing
with several countries. In fact, even national definitions fluctuate according to the stage of
development, as exemplified by changes in the US Standard Metropolitan Area definition
in the 1970s, 'because society continues to change, criteria and areas (concerned) which at
one period of time may have been valid representations of conditions and concepts cease to
be so with the passage of time' (Berry *et al.* 1968). Thus, the quest is to provide a general
working definition of urban reality which is capable of transcending space and time.

Bearing in mind the need to ascribe population to space, a range of administrative definitions and associated delimitation rules have been defined in various countries, to locate
and count people and their activities. Nevertheless, several important problems have yet to
be overcome. If an agglomeration of population is a specified contiguous space, how might
it be recognized? In delineating the urban/non-urban dichotomy at the urban fringe, where
should a boundary be drawn so as to represent realistically the functional relationships with
other settlements in the immediate hinterland? Aerial photographs have been used in several
countries in an attempt to answer the first question, and today other forms of remotely sensed
data are also being used successfully. With regard to the second question, while interpreters
of aerial photography have been used to delimit urban areas, digital remote sensing systems
make it possible to assess contiguity automatically, in accordance with a range of criteria.

8.2   Defining Urban Agglomerations

As has been noted, definitions are not ends in themselves. In fact, two kinds of argument
are usually put forward in this context: the rationality of economic reasons and the need to
translate the agglomeration concept into statistical practice.

---

[1]The concept of the agglomeration will be analysed here, rather than that of city-centre, because it better
corresponds to the reality of a continuum of populated nuclei comprising part of a single or multiple administrative
authorities. Thus, it not only takes into account population size, but also the characteristics of the associated area.
Such a focus implies a spatial hierarchy of urbanized areas.

There are implications of membership (or otherwise) of particular city networks, such as 'one of the top ten cities' in the world or in a continent. These labels connote the 'image' of a given agglomeration and have associated economic and political spin-offs in terms of, for example, investment by international companies. Today, despite the resistance of national administrations and administrative obstacles, there is a world order of economic and political networks. Consequently, regions are developing outside the traditional conceptions of networks and hierarchies. For instance, being a part of the 'European Blue Banana' (a region which stretches from Milan to Barcelona through Brussels and the Rühr Area) is considered by some of the member cities to be a bonus in terms of the image they wish to project. In a similar vein, the Atlantic cities have the analogous club, known as the 'Atlantic Arc'. So being 'in' connotes dynamism, while being 'out' may be risky and lead to being marginalized — excluded from the mainstream!

But even within such regions the relations between the centre and the periphery differentiate between a dynamic 'leader engine' and others. Each agglomeration seeks to be considered the leader in order to attract maximum investment (economic, technological, social, research, etc.). Leadership has implications for the central agglomeration, but different (positive and negative) effects for the rest of the territorial entity. Whatever the specifics of particular situations, relations between cities become one of the major imperatives for integration and dynamism within urban regions: 'In Europe, 120 agglomerations concentrate more than half the population, but more than just people, they concentrate innovative capacities; it is the diffusion of these innovations and development opportunities which affects their hinterlands and, consequently, the European Community as a whole' (free translation from Maragall i Mira, 1992).

8.2.1 Definition Criteria

Thus, there is a need to develop criteria which foster comparison and allow group membership to be ascertained. Classifications are not innocent and results are often embedded in the specification of measurement criteria. The criteria used might change the details of the city rankings, while the thresholds used might be manipulated to the same end. To avoid any doubt or speculation, the concept of an agglomeration must be clearly defined using consistent, incontrovertible indicators pertaining to network relations and other statistical measures. Such a description has to be as complete a representation as possible of the urban reality. In order to handle complex concepts, elements that can be easily manipulated need to be identified. Moreover, it is important that the constituent indicators are semantically unambiguous, and can be collated with approximately the same level of confidence in different countries. Unfortunately, this is seldom achievable, as has been noted in several studies (OCDE 1988, ERIPLAN 1975, cited in Pumain *et al.* 1992). For instance, cities may be compared using a wide range of criteria, such as population, area, activities, economic trends or functions, or journey to work characteristics. But even these apparently straightforward criteria are not really precise:

- the term agglomeration implies a population concentration which implies a minimum area and an associated density threshold;

- the range of activities is linked to (un)employment rate, population, commuting, *etc*; and

- the size of territory is associated with continuity between separate urbanized areas.

Each of these criteria is invariably delicate to handle, reflecting different realities beneath near identical definitions, and despite numerous convergences (Pumain *et al.* 1992). Thus, it is necessary to devise the simplest possible definitions in the interests of semantic homogeneity, and the best possible understanding of the criteria employed. These issues are particularly important to promote common and widespread understanding of the underlying concepts, because behind struggles for primacy 'the need for comparable statistics is sought by European entities looking for European equilibrium, by states charged with territorial planning, and most of all by administrative urban units that wish to order themselves among European partners' (free translation from Pumain *et al.* 1992). This is now of major interest in a continent with open boundaries, where subsidiarity is of prime importance.

In considering issues of homogeneity and transparency in the definition of agglomerations, two points are worthy of further treatment:

- if the urban phenomenon is characterized by a rural-urban continuum, it is necessary to develop a precise and clear definition of land use and land cover categories so as to be able to define the limit of urban areas;

- as urban agglomerations represent a common spatial form of urbanization in Europe, the morphology of urban areas might be considered to be the most objective and easily-obtained criterion for defining contiguous built up areas.

Given recent technological advances in computer mapping, aerial photography and remote sensing, we might reasonably anticipate that analysis of digital remotely-sensed data holds the key to developing appropriate indicators of urban agglomeration. Specifically, remote sensing might be used to:

- identify characteristic spatial features; and to

- help test and assess the criteria and thresholds to be agreed upon for a common understanding of urban agglomerations.

8.3 Spatial Forms — Urban Forms?

Given that the form of the city is unlikely to represent its underlying structure, it is really rather limiting to look only at aspects of the surface expression of urban land cover and land use. Such analysis ignores the diversity of socio-economic, cultural and political interactions that take place within urban areas, the existence of technical and facilities networks, and the reality of more intangible links (e.g. decision networks and information networks). Nevertheless, today's advanced technologies provide the opportunity to take into account, through emergent 'spatial forms'[2], some elements from natural landscapes and from human

---

[2]The term 'spatial forms' refers an object or set of surface elements which is detected as a whole.

activities. Consequently, the spatial forms detected and identified might be considered not just as the outcome of human interactions in space, but rather as revealing socio-economic and political trends over time.

Remotely sensed data generate products which can help to meet the information needs of urban analysis. High spatial resolution, Earth-orbiting satellite sensors, in particular, are becoming an important source of spatial and temporal information on urban areas.

8.3.1 The Relationships Between Urban Forms and Remote Sensing Data

What concepts are associated with the word 'form'? Referring only to the level of observation mentioned above — i.e. the spatially identifiable elements — it seems reasonable to observe that the discernibility of an object depends on the difference between it and its surroundings, on the contrast between the object and its background. In short, an object appears through apparent order amongst more general disorder. More generally, the notion of 'form' involves two distinct spatial reference systems:

1. Cartesian space, with distances and quantitative results;

2. Representative space, without mass, speed or size, discernible through signs and signals — an ordering of qualitative discontinuities on a spatial frame.

These two reference systems are 'both essential and contradictory; irreducible one to the other' (Moles and Rohmer 1972). This notion is interesting in the context of remote sensing because both aspects are present within an image: the measurements and their representation create the qualitative information about land cover or land use.

The examination of urban form is a way of contemplating the evolution of urban society, and the way in which changes in societal structure over time are manifest spatially. Remote sensing data provide a useful mechanism through which these forms can be appreciated, even if, as stated earlier, the signals received by the remote sensing device cannot be linked directly to all properties of interest pertaining to the object of study. The information produced by remote sensing is spatially referenced through an implicit geometric location of the pixels. Repeated observation of the same area on the Earth surface gives the opportunity to update this information over time. Various urban forms are potentially discernible using such devices, including linear objects (e.g. roads and rivers), although small features cannot always be accurately discerned (Dowman and Peacegood 1988). However, it should be noted that the relationships between urban surface elements and the radiance/reflectance values recorded for each pixel (picture element) in the image are not straightforward. Moreover, the ability to distinguish different surface features will vary according to the sensor spatial resolution, and is dependent on the spatial and spectral heterogeneity of the urban surface cover. Thus, the results obtained might differ between countries, depending on the characteristic urban structure in each: in simple terms, European, Asian and African cities will not be classified into the same categories. The cultural and historical bases (e.g. building habits and materials, and planning rules) and the environmental surroundings (e.g. climate and vegetation) will introduce variations into the results of such image processing. Finally, it is noted that the structures emerging through an analysis of urban morphology are associated with particular cultural artifacts, and will appear or vanish at different spatial scales. Multi-scale analyses can help to highlight these effects, elucidating the processes that led to their formation.

8.3.2   Statistical Information Needs and Remotely Sensed Data

It has been stressed previously that the extent of built-up areas might be an appropriate indicator of urban population density and urban shape continuity, and that this might be central to creation of a benchmark definition of 'urban agglomerations'. Clearly, any such definition must be both simple and objective.

The general urban form, sometimes referred to as the Morphological Urban Area (MUA), could provide a measurement of the compactness of the urban space and the 'holes' in the urban fabric, as well as a benchmark against which urban growth might be measured. The MUA concept (Eurostat 1994) has been used in several European countries (*inter alia* Belgium, Denmark, Spain, France and Greece) and the experience of its usage has underscored the need for benchmark standards.

Because, in the context of this Chapter, the concept of agglomeration is primarily concerned with the contiguity of built-up areas, it is absolutely necessary to:

1. clarify the definition used to differentiate urban and non-urban land; and

2. determine the minimum population threshold above which a human settlement may be considered to be an urban area.

The second part of the definition brings into play considerations of the minimum distance between settlement nuclei separated by rural land. In France, as in Denmark and Ireland, settlements separated by a distance of more than 200m are considered to be distinct spatial entities. Superimposed on this is a further criterion concerning the minimum size of population for a settlement to be considered, which varies from one country to another (50 in France, 200 in Belgium, 500 in Scotland). Remote sensing can provide quantitative and repeatable measurements of the first of these properties, and can be used to make estimates of the second. It can therefore make an important contribution to the definition and delineation of urban entities. It also makes it possible to define the scope and limits according to scenarios which use:

- different urban indicators (e.g. with or without structuring networks, with or without industrial and commercial areas, and with or without scattered settlements); and

- different spatial structures (e.g. the built versus non-built dichotomy, or other urban land use categories).

A number of studies has been conducted, initially under the auspices of Eurostat (the Statistical Office of the European Communities), using remotely-sensed data to map and monitor urban areas (Eurostat 1994, Eurostat 1995, Eurostat 1998). Each used a slightly different approach (Eurostat 1993, Cauvin 1994, Donnay 1994, Eurostat 1994, Weber *et al.* 1995, Barnsley and Barr 1996, see also Barnsley *et al.*, Chapter 7). In general, however, the common objectives have been to determine the limits of the MUA and to compare the results obtained from remote sensing methods with official statistical definitions. Most of the experiments have focussed on deriving information on urban land use (c.f. land cover). The land use categories were developed from the CORINE (COordination et Recherche d'INformations sur l'Environnement) Land Cover classification in a scheme that has become known as CLUSTERS (Classification for Land Use STatistics, Eurostat Remote Sensing project; Table 8.1). One of the main problems with nomenclature schemes such as these, is that concepts of land use and land cover are frequently mixed or, worse, used interchangeably. In reality, they have very different connotations. The problem is compounded by the difficulty of determining either land cover or land use unambiguously using remotely-sensed images, which merely record the levels of radiation reflected, emitted or scattered from the Earth surface. One solution is to introduce ancillary spatial data by means of a Geographical Information Systems (GIS). Of course, even this does not remove all of the problems associated with delineating the zone of different urban land use or risk of different implementations by image interpreters/analysts from separate countries.

8.4 Application and Discussion

8.4.1 Application

It is interesting to develop the conceptual approach outlined above in the context of an empirical study of the Strasbourg 'agglomeration'. The nature of the data used conditions the delimitation of the urban morphological area. Thus, the first experiment was carried out using a SPOT-HRV multispectral image (acquisition date — 28/06/1986; scene number — KJ 50-252; processing level — Level 2B), while the second was performed using a SPOT-HRV Panchromatic image (acquisition date — 15/09/1986; scene number — KJ 50-252; processing level — Level 2B). Different image processing routines were applied to each image:

- the multispectral image was processed using classification/fusion techniques, in which the results of a conventional, per-pixel classification were combined with discrete zonation data in order to minimize classification errors. The urban area was delineated on the basis of land cover, splitting various built categories from vegetation and water;

- the panchromatic image was processed using a texture filter (specifically, a 7 × 7 circular operator) with the aim of identifying the built-up zones.

The total extent of the Strasbourg agglomeration was then identified by combining the data derived from each of these images and recoding them into a binary 'urban' vs. 'non-urban' image. Further manipulations were applied to each image based on the distance and population thresholds outlined in the previous section, namely:

- a population criterion (a minimum population size of 2000), and

- a distance criterion (a minimum distance of 200m between settlements to be grouped into a single agglomeration).

Table 8.1: CLUSTERS nomenclature (artificial surfaces).

Level I	Level II		Level III		Level IV	
Artificial surfaces	A1	Urban fabric	A11	Residential zones	A111	Continuous and dense residential areas
					A112	Continuous residential areas
					A113	Residential surface of suburban type
					A114	Discontinuous residential areas
					A115	Collective residential
			A12	Public services and local government	A12	Services
	A2	Industrial and commercial activities	A20	Industrial and commercial activities	A201	Activities of heavy industry
					A202	Other industrial activities
					A203	Commercial/financial services/activities
					A204	Agricultural holdings
	A3	Technical, transport and communication infrastructure	A31	Technical infrastructure	A311	Technical networks & protective structures
					A312	Traitement des eaux
			A32	Transport and communication	A321	Transport and communication road
					A322	Rail networks
					A323	Airports and aerodromes
	A4	Extractive industries, building sites, tips and waste land	A41	Extractive industries		
			A42	Building sites, tips and waste land	A421	Building sites
					A422	Tips
					A423	Waste land
	A5	Land developed for recreational purposes	A50	Land developed for recreational purposes	A501	Historic sites
					A502	Sports facilities
					A503	Other developed areas

These were applied in several different ways to decide which of the surrounding settlements should be incorporated into the Strasbourg agglomeration (Figure 8.1).

It is interesting to evaluate the differences between the results obtained using different combinations of the image data and criteria outlined above. Figures 8.2 and 8.3, in which urban land is shown in black, are derived from the classified multispectral image. In the first of these (Figure 8.2), all of the buildings are kept, while in the second (Figure 8.3) only the inhabited buildings are extracted. A buffer zone of 200m has been created around the derived urban areas to derive the MUA. It is worth noting that roads and railway infrastructure are included in the urban structure by the image classification process. The resultant urban agglomerations exhibit different degrees of compactness: the second image (Figure 8.3) displays a smoother boundary and implies a more distributed population. The third image (Figure 8.4) is derived from the panchromatic image alone. The resultant urban agglomeration is still more compact than either of the two previous examples. Moreover, some of the 'holes' apparent in the urban fabric of Figures 8.2 and 8.3 are no longer present.

8.4.2 Discussion

Of course the results presented above are indicative rather than definitive, and the final delineation and definitions have in any case to be made by the national statistical offices. But the approach developed in this paper is instructive, in that it clearly demonstrates that there is no unique solution which yields an objective and unambiguous delineation of the morphological urban area. Even if remote sensing offers interesting solutions, the rules must be precisely defined:

1. using the classification nomenclature, careful interpretation of urban limits has to be set up to avoid human errors, and if an automatic approach (as opposed to manual/visual interpretation) is used then the selected urban classes must be clearly specified; and

2. mathematical morphological processing permits comparison between different geographical schemes.

The aggregation criteria (e.g. the distance threshold, the importance assigned to transport infrastructure and the basic population settlement criterion) must be also tested in various locations in order to find the best fit to reality.

8.5 Conclusion

This Chapter has examined the potential use of satellite remote sensing for deriving information on a consistent measure of the extent of urban areas, namely the Morphological Urban Area (MUA). A wide range of digital image-processing techniques can be applied to remotely-sensed data to yield information on the MUA, of which this Chapter has considered but a few. Ultimately, we must test a number of other approaches, evaluating their accuracy and reliability. Whichever techniques prove most successful in the long run, the development of a consistent, objective and accurate approach to determining urban agglomerations from satellite sensor data is an important aim, offering improved knowledge of the

154

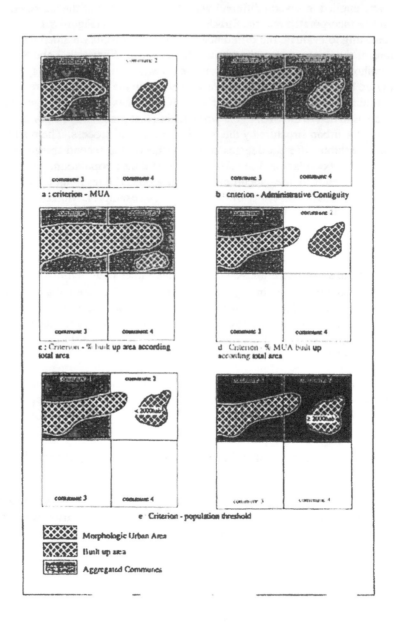

Figure 8.1: Multi-criteria approach to urban area delimitation: (a) Morphological urban area (MUA), (b) adding contiguous administrative areas, (c) built-up area as a percentage of total area, (d) MUA as a percentage of total area, and (e) application of population thresholds of greater than and less than 2000 residents.

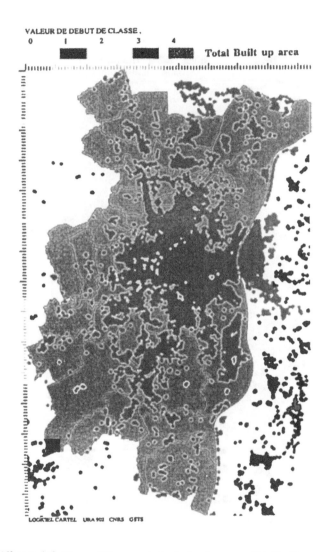

VALEUR DE DEBUT DE CLASSE.

0   1   2   3   4

Total Built up area

LOGICIEL CARTEL   URA 902   CNRS   GSTS

Figure 8.2: Different definitions of the morphological urban area: Total built up area. The image covers the Communauté urbaine de Strasbourg and is based on a SPOT-HRV multispectral image, © CNES/SPOT IMAGE.

Figure 8.3: Different definitions of the morphological urban area: Inhabited built up area. The image covers the Communauté urbaine de Strasbourg is based on a SPOT-HRV multispectral image, © CNES/SPOT IMAGE.

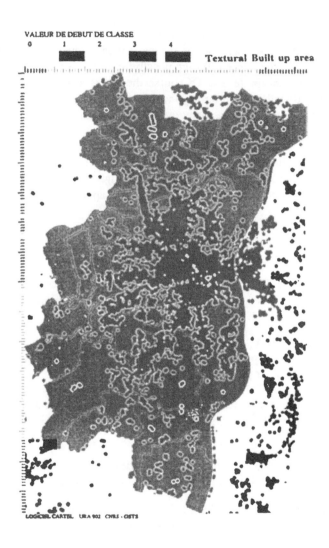

Figure 8.4: Different definitions of the morphological urban area: Textural built up area. The image covers the Communauté urbaine de Strasbourg is based on a SPOT-HRV multispectral image, © CNES/SPOT IMAGE.

current state and rates of development of urban areas and providing comparable statistical information between countries and across different time periods. The approach has particular potential for the delineation of urban areas in developing countries, where the mapping base is frequently of lower quality and is often dated (see also Baudot, Chapter 12 this volume). Nowhere is this more important than in the mapping and monitoring of informal settlements which are often overlooked in official documents.

Remotely-sensed data also facilitate the determination of the spatial composition and orderings of urban areas and hence offer the possibility to focus on issues of urban management at a range of decision levels. Of course the stakes associated with the need for comparable urban information reach far beyond this and, even if sound criteria can be promoted through remotely sensed data, many questions remain: What is urban? Are urban limits sharp lines or fuzzy zones? What happens in the cases of conurbations or polynuclei settlements? Looking at the urban forms generated from satellite sensor images suggests a passage from continuity to discontinuity. Ultimately, image processing needs to be regarded as one possible approach to identifying urban form, although the validity of the different scenarios has to be tested in order to select the best possible approach for a range of possible applications.

8.6   References

Barnsley, M. J., and Barr, S. L., 1996, Inferring urban land use from satellite sensor images using kernel-based spatial reclassification. *Photogrammetric Engineering and Remote Sensing*, 62, 949–958.

Berry, B. L. J., Goheen, P. G., and Golstein, H., 1968, Metropolitan Area Classification: A Review of Current Practice, Criticisms, and Proposed Alternatives, Working Paper 28, U.S. Bureau of the Census, Washington D.C.

Cauvin, C., 1994, Densités urbaines. Bâtis et populations: modeles spatiaux, images satellitaires et représentations, Scientific Report URA 902 CNRS, Laboratoire Image et Ville, Strasbourg.

Donnay, J.-P., 1994, Agglomérations morphologiques et fonctionnelles, l'apport de la télédétection urbaine. *Acta Geographica Lovaniensia*, 34, 191–199.

Dowman, I. J., and Peacegood, G., 1988, Information content of high resolution satellite imagery. *International Archives of Photogrammetry and Remote Sensing*, 27, 84–93.

ERIPLAN, 1975, Etude prospective d'amégement et d'environnement dans la mégalopole en formation de l'Europe du Nord-Ouest, Report to the european communities (4 volumes), ERIPLAN, The Hague.

Eurostat, 1993, *The Impact of Remote Sensing on the European Statistical System — Proceedings of the Seminar, Bad Neuenahr* (Luxembourg: Office for Official Publications of the European Communities).

Eurostat, 1994, *Pilot Project Delimitation of Urban Agglomerations by Remote Sensing: Technical Report* (Luxembourg: Office for Official Publications of the European Communities).

Eurostat, 1995, *Pilot Project Delimitation of Urban Agglomerations by Remote Sensing: Results and Conclusions* (Luxembourg: Office for Official Publications of the European Communities).

Eurostat, 1998, *The Impact of Remote Sensing on the European Statistical System — Proceedings of the Seminar, Esquilino, Rome* (Luxembourg: Office for Official Publications of the European Communities).

Maragall i Mira, P., 1992, La ville dans l'Europe. *Territoires*, 327, 46–51.

Moles, A., and Rohmer, E., 1972, *Psychologie de l'espace* (Paris: Casterman).

OCDE, 1988, Statistiques urbaines dans les pays de l'OCDE, Programme des affaires urbaines, OCDE.

Pumain, D., Saint-Julien, T., Cattan, N., and Rozenblat, C., 1992, Le concept statistique de la ville en Europe, In *Collection Eurostat — Th'eme 3 — Série E* (Luxembourg: Office des Publications Officielles des Communautés Européennes).

Weber, C., Hirsh, J., and Serradj, A., 1995, Pour une autre approche de la délimitation urbaine, In *Actes des Sixièmes Journées du Réseau de Télédétection de l'UREF: Télédétection des Milieux Urbaines et Périurbaines*, UREF (Liege: UREF).

PART IV

DEFINING URBAN POPULATIONS OVER SPACE AND TIME

CHAPTER 9

Measuring Urban Morphology using Remotely-Sensed Imagery

Paul A. Longley and Victor Mesev

9.1 Introduction

Previous chapters have considered how detailed urban morphologies might be revealed using satellite technology. The focus has been upon quite general-purpose technical solutions to technical problems, and has been couched largely within the physical domain of the built environment. This Chapter addresses the applicability of some of this technical work to the derivation of summary indicators of urban shape and form. In so doing, we are minded to begin by stating two important considerations. The first is that cities are socio-economic systems, and hence 'urban analysis' is at least as much about human activity patterns (journeys to work, recreation, shopping behaviour, residential differentiation, etc.) as it is about the built environment. The built environment is a frozen artefact of the concatenation of past and contemporary urban processes (see also Batty and Howes, Chapter 10 this volume), and reveals but one facet of the functioning of urban systems. The second derives directly from this, and is that the power of even the most sophisticated optical satellite sensors to discriminate detailed urban land-use categories is severely limited (see also Donnay et al., Chapter 1 this volume). The OrbView 3 (OrbImage), QuickBird 1 (EarthWatch) and IKONOS 1 (Space Imaging) satellite sensors will provide unprecedented spatial information about the Earth's surface, at a level of detail ($1m$–$5m$) more closely associated with aerial photography. Even at these fine scales, however, it will not be possible unambiguously to identify, say, small workshops amidst residential areas from an analysis of their spatial form alone. In any case, a much more time- and cost-effective route is likely to involve interfacing of socio-economic information with data pertaining to the socio-economic *function* of the built *form*. There is thus a clear substantive and methodological rationale for integrating socio-economic information with satellite imagery.

Yet integrated approaches, such as this, will never offer a universal panacea for urban analysis. Indeed, in adding a new 'urban' focus to remote sensing research, the objectives of measurement will undoubtedly change and diversify. The urban focus presents much more than technological challenges, and the remote sensing community must heed the tide of relativism that has run through the social sciences during the last 25 years. The traditional remit of remote sensing (e.g. land inventory analysis and natural resource assessment) is to resolve essentially single criterion problems with 'objective' solutions. By contrast, the domain of planning and urban analysis is characterized by multi-criteria decision-making, which is typically very sensitive to context. Nevertheless, the prospects for harnessing remote sensing technologies to such tasks are bright. There is a huge range of digital socio-economic data that is now available for use alongside satellite imagery (Mesev *et al.* 1996). Moreover, there are real prospects for creating far better data models of socio-economic environments, as improved remotely-sensed data products come on stream. The applicability, and ultimate application, of such models will, however, require the adoption of a 'horses for courses' approach to data modelling, for there is no objective, universally recognized social reality, and the depiction of social realities firmly defines the scope for their subsequent analysis. With these considerations in mind, this Chapter suggests various ways in which remotely-sensed data might inform urban theory and urban planning policy, and begins to assess the prospects for using socio-economic data to foster progress towards these goals.

9.2   From Data-Led Theory to Theory-Led Data Analysis?

In an excellent guide to the development of socio-economic Geographical Information Systems (GIS), Martin (1996) describes how the handling of socio-economic information within GIS has developed from its roots in remote sensing and computer-assisted cartography in an *ad hoc* and data-led manner.[1] A particular focus has been upon the ways in which data transformations are invoked in the collection, input, storage, manipulation and analysis of spatial data. Remote sensing has never been beset with problems of the quantity of data, and the primary goal of most image classification exercises has been to structure and classify large volumes of potential information. As will been seen below, this state-of-affairs is a much more recent development in the analysis of socio-economic systems, and the effective management and analysis of geographical information is a recurrent challenge of the late 1990s. Considerable progress has been made with technical problems which have generic solutions, such as vector-to-raster conversion, but other problems — notably the definition and analysis of socio-economic entities — are likely to require much more context-sensitive and customized solutions. This is particularly likely in the context of development of pan-European analysis, given that the areal entities used to structure socio-economic data are cast within the different administrative geographies of individual nation states.

Remote sensing has always been a data-rich subject, and yet irrespective of increases in satellite precision, improvements in the detection of built *form* will provide only a limited perspective upon the *functioning* of urban settlements. By contrast, socio-economic information almost invariably provides an imprecise representation of form (e.g. because of confidentiality restrictions on the availability of census data), yet it tells us rather more about the activity patterns which characterize the functioning of settlements. If the push

---

[1] For a more general statement of the relations between data models and data products, see Davis *et al.* (1991).

towards ever-improving satellite sensor resolutions is ultimately unlikely to reveal the form and function of individual elements of the urban mosaic, so different aggregation difficulties are likely to be removed from the analysis of socio-economic distributions. The end result is nevertheless the same — namely, a reliance upon modelling the spatial distributions of 'geographical individuals' within areal entities. GIS provides a suitable environment in which to assess the impact of different zonations (Openshaw 1996), different area-preserving and/or volume-preserving transformations (Martin 1996), error modelling (Goodchild *et al.* 1992) and the compounding of errors in overlay analyses. The sources and operation of errors in urban remote sensing are likely to be different from those arising in socio-economic GIS, and thus analysis of the two alongside one another is likely to lead to the development of less error-prone models of reality.

This is part of a wider agenda for the integration of spatial analysis into GIS in both the academic (Fotheringham and Rogerson 1993) and applied (Clarke and Clarke 1995) realms, a phenomenon which is likely to transpire in any case as GIS becomes an all-pervasive 'background' technology for geographical information handling (Longley and Batty 1996). An implication of this is that GIS-based urban analysis will develop upon the foundations of data models, and that the distinction between data and other analytical models is artificial, but blurred. Each develops through incremental procedures of inductive generalization and deductive spatial forecasting, using appropriate combinations of generic and context-sensitive techniques for information handling.

A necessary precursor to the analysis of urban morphology will be a fuller and more holistic grasp of the aspects of urban morphology with which we wish to be concerned — urban 'morphology' variously implies 'form', 'land use', and 'density', and each of these, in turn, has connotations with the shape, configuration, structure, pattern and organization of land use, and the system of relations between them (Whyte 1968, Batty and Longley 1994, Barnsley *et al.* this volume). In the context of remote sensing of human settlements, concern with land use has focused upon the central 'urban/non-urban' dichotomy. This is most clearly manifest in physical form, although in practice finer disaggregations, as well as measures of the intensity of human activities, are also desirable. The density of a particular land use may be couched in terms of the physical presence or absence of a particular land use, while human activity-based conceptions of urban land use require closer attention to residential densities.

If the conception of a phenomenon prescribes the way in which it is measured and this, in turn, prescribes the framework for analysis, then it is useful develop a framework which is consistent through all stages of this process. It is our contention that fractal geometry provides such a framework (Batty and Longley 1994). In essence, fractal geometry posits that the way in which space is filled may be conceptualized as a continuum of dimensions, with the Euclidean dimensions of 0 (a point), 1 (a line), 2 (a plane) and 3 (a solid) being only special cases. Fractal shapes also exhibit self-similarity (or self-affinity) across a range of scales, and this property has been detected using a variety of statistical methods (Goodchild and Mark 1987, Batty and Longley 1994). Human geographical phenomena have long been conceived as exhibiting properties of such recursive self-similarity, with central place theory perhaps providing the best known example. Within Geography, however, the relevance of central place theory is restricted to the domain of pedagogy, and there have been few systematic attempts to define the structured ordering of urban space since the 1960s — when data sources and measurement technologies were very much cruder. Much the same is true for other measures of size and shape (Haggett *et al.* 1977) and studies of the 'allometric' scaling between the population size and area of urban settlements (Longley *et al.* 1991). The last 30 years has seen a gradual but cumulative loosening of the controls on European urban development, for example, and planning has responded to the changing political economy of spatial development by becoming much less deductive and less normative in approach (Batty 1995). We have argued elsewhere (Mesev *et al.* 1995) that fractal geometry provides a flexible yet theoretically consistent means of characterizing the morphologies of different urban settlements across space and time.

9.3   Morphological Analysis of the 'Fractal City'

We have alluded to the need to devise application-sensitive digital data sets, and this becomes apparent if we consider the wide range of reasons why digital urban data sets are of central importance to urban analysis. Urban theory has variously considered densities of population and densities of economic activity across urban areas, to devise measures of the efficiency of urban form in terms such as transportation (Benguigui and Daoud 1991), energy efficiency (Rickaby 1987), residential development, retail and industrial planning, and access to public open space (Longley *et al.* 1992). Repeated measurements of urban morphology across different time periods can generate insights into the kaleidescope of change in land use patterns. Remote sensing can provide a convenient estimate of the distributions and densities of these activities, which are transferable across space and time. Study of the physical location and spatial patterning of built forms makes possible analysis of cities as networks and dynamic entities, thereby forging links between the spatial form of any settlement and its functioning (Batty and Longley 1994).

Table 9.1 records some of the different urban objects that have been subjected to fractal analysis, together with their associated empirical measures (all results are reported to the number of decimal places published and, where there are several different estimates, in particular from Batty and Longley and from Frankhauser, the lower estimates of fractal dimension, D, are given). Fractal properties have been identified for quite diverse properties of urban areas. Although many of these applications predate the use of digital data for morphometric analysis, the range of quite diverse urban applications indicates that the corresponding digital databases may need to be application-specific. The second point to emerge from Table 9.1 is that the measures do not exhibit more than a broad consistency of numerical ordering. As in the earlier work on urban allometry, these divergent results must in part reflect the range of different data sources used, since comparable measurements are unlikely to arise when they are not based upon comparable datasets. This underscores the importance of ESF initiatives such as GISDATA in standardizing the collection and input of digital data into GIS: even if the history of GIS has been about data leading theory (Martin 1996), there are nevertheless many areas of urban analysis in which urban theory requires improved data for robust and defensible empirical application. Such consistency will enable comparative urban morphologies to be investigated across a range of scales: at the macro-scale, entailing the comparison of the overall morphology of different settlements; at the meso-scale, focusing upon the morphological make-up of different areas of single settle-

Table 9.1: The preliminary evidence for fractal cities; adapted from Batty and Longley (1994). Key: B&D — Benguigui and Daoud (1991); B&L — Batty and Longley 1994; Dox — Doxiadis (1968); Fra — Frankhauser (1988, 1990, 1992, 1993); T&M — Thibault and Marchand (1987); Dox/Ab — compilation of data from Doxiadis (1968) and Abercrombie (1945); Sm — Smith (1991).

Settlement	D	Settlement	D
Urban Development Patterns		**Urban Growth Patterns**	
Albany 1990 (B&L)	1.494	London 1820 (Dox/Ab)	1.322
Beijing 1981 (Fra)	1.93	London 1840 (Dox/Ab)	1.585
Berlin 1980 (Fra)	1.73	London 1860 (Dox/Ab)	1.415
Boston 1981 (Fra)	1.69	London 1880 (Dox/Ab)	1.700
Budapest 1981 (Fra)	1.72	London 1900 (Dox/Ab)	1.737
Buffalo 1990 (B&L)	1.729	London 1914 (Dox/Ab)	1.765
Cardiff 1981 (B&L)	1.586	London 1939 (Dox/Ab)	1.791
Cleveland 1990 (B&L)	1.732	London 1962 (Dox/Ab)	1.774
Columbus 1990 (B&L)	1.808		
Essen 1981 (Fra)	1.81	Berlin 1875 (Fra)	1.43
Guatemala 1990 (Sm)	1.702	Berlin 1920 (Fra)	1.54
London 1960 (Dox)	1.774	Berlin 1945 (Fra)	1.69
London 1981 (Fra)	1.72		
Los Angeles 1981 (Fra)	1.93	**Transport Networks**	
Melbourne 1981 (Fra)	1.85		
Mexico City 1981 (Fra)	1.76	*Suburban Rail*	
Moscow 1981 (Fra)	1.60	Lyon I 1987 (T&M)	1.88
New York 1960 (Dox)	1.710	Lyon II 1987 (T&M)	1.655
Paris 1960 (Dox)	1.862	Lyon III 1987 (T&M)	1.64
Paris 1981 (Fra)	1.66	Paris 1989 (B&D)	1.466
Pittsburgh 1981 (Fra)	1.59	Stuttgart 1988 (Fra)	1.58
Pittsburgh 1990 (B&L)	1.775		
Potsdam 1945 (Fra)	1.88	*Public Bus*	
Rome 1981 (Fra)	1.69	Lyon I 1987 (T&M)	1.45
Seoul 1981 (B&L)	1.682	Lyon II 1987 (T&M)	1.00
Stuttgart 1981 (Fra)	1.41	Lyon III 1987 (T&M)	1.09
Sydney 1981 (Fra)	1.82		
Syracuse 1990 (B&L)	1.438	*Drainage Utilities*	
Taipei 1981 (Fra)	1.39	Lyon I 1987 (T&M)	1.79
Taunton 1981 (B&L)	1.636	Lyon II 1987 (T&M)	1.30
Tokyo 1960 (Dox)	1.312	Lyon III 1987 (T&M)	1.21

ments; and at the micro-scale, in which the effect of individual activity patterns upon the make-up of urban areas might be investigated (Orford 1999). Fractal geometry provides a potentially unifying framework for the analysis of urban phenomena across all of these scales. In our empirical case study, we develop applications at the macro- and meso-scales to investigate the fractal morphology of urban land use as a blanket category, as well as the morphologies of constituent land-use types.

One further important repercussion of the innovation of fractal geometry has been the derivation of space-filling 'norms' against which the density gradients of real-world cities can be compared. For example, Batty *et al.* (1989) describe how the accretive structures generated by diffusion-limited aggregation (DLA) provide a fractal 'benchmark' for the dimension of real-world structures. Fotheringham *et al.* (1989) use this benchmark to challenge the widely-held notion that urban density gradients characteristically decline over time, arguing that the imprecise measurements of previous research do not accurately portray the dynamics of growth along the outer edge, or 'growth zone' of urban settlements. This remains to be verified through detailed empirical analysis, and use of satellite imagery presents a useful way around the pervasive problems of under-bounding and over-bounding in urban analysis. Appropriate measurement techniques, harnessed to 'good' and appropriate databases will provide a means of developing our measures of urban sustainability (Rickaby 1987) and the efficiency of urban forms, using a range of measures consistent with understanding built forms and socio-economic distributions.

Measurements may often be only crude, but they are nevertheless required if we are to produce human activity-related measures of form which are relevant to urban policy. The spirit of the approach outlined in Mesev *et al.* (Chapter 5, this volume) is that the estimates of population and residential densities obtained from remote sensing sources are essentially 'first estimates', which may be further refined through the integration of socio-economic data sources into land-use classification algorithms (Mesev 1998, Baudot this volume). Indeed, a range of diverse urban data models may now be combined into integrated data sets which are usable in the measurement and classification of urban settlements, and the prospects are now bright for the development of spatially-extensive models of urban structure.

9.4   Some Measures of Relevance to Urban Morphology Research

A recurrent theme in the study of the growth and evolution of human settlement systems has been the ways various attributes of such systems scale with increasing size: the study of relative sizes being allometry (Gould 1966). This research derives from the locational analysis tradition in human geography (Haggett *et al.* 1977) and, as such, predates the innovation of remote sensing technology to generate detailed measurements of urban land cover. It is only now, following the development of enhanced image analysis of urban areas (Barnsley *et al.* 1993, Mesev 1998, Mesev and Longley 1999), that the size and density characteristics of urban areas may be investigated in a more precise way.

The two basic measures of size are population and area, both of which may be imputed using remotely-sensed data. Associated with the population, N_k, of any urban cluster, k, there is a number of ways in which area (and hence density) might be defined. Batty and

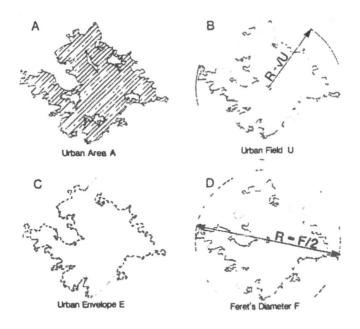

Figure 9.1: Measures of urban morphology: a) urban area; b) urban field; c) urban envelope; d) radius.

Longley (1994) develop two measures of area in their preliminary analysis of the size, shape, scale and fractal dimension of urban areas: first, there is the occupied area, A_k, which can be defined loosely as the built-up or developed area, and is likely to co-vary to an extent with population. Second, there is the urban field whose area, U_k, can be defined as the hinterland immediately associated with the greatest radial extent of the urban cluster (Hägerstrand 1952): this may be, for example, the immediate circle within which growth has already taken place. There is also a further variable of interest which relates area, A_k, to field size, U_k; this is the urban envelope, E_k, defined as the length of the boundary or perimeter which marks the greatest extent of the built-up area. These measures are illustrated in Figure 9.1.

Figure 9.1a shows the built-up urban area the extent of which, A_k, is indicated by the cross hatch. It is this area that contains the population, N_k. The urban field is shown in Figure 9.1b. This is the bounding circle, based on the centre of the cluster, which is marked by the maximum radius, R_k, which contains the whole cluster. The area of the cluster is given as $U_k = \pi R_k^2$, where $U_k > A_k$. The urban envelope is shown in Figure 9.1c, its length, E_k, being a measure of both the size and the shape of the cluster. Figure 9.1d shows the maximum spanning distance across the cluster: the length of this span is defined as F_k. This measure has been used by Batty and Longley (1994) to approximate the radius of a settlement, R_k. Two types of relationship between these variables may be investigated: first, relating population, N_k, to area, A_k, and to field radius, R_k; second, relating the length of the envelope, E_k, to these same variables. These types of relationship are central to allometry or 'relative size' relationships (Gould 1966). By relating size and length to area, Batty and Longley (1994) were able to explore questions of density and to relate this work to the mainstream literature on urban allometry (Dutton 1973).

The classical allometric relation between population size, N_k, and occupied area, A_k, is:

$$N_k = \gamma A_k^\beta = \gamma A_k^{\frac{\Delta}{2}} \tag{9.1}$$

where

γ is a constant of proportionality and

β is a scaling constant.

In this equation, β has also been written as $\frac{\Delta}{2}$ where Δ can be interpreted as a 'dimension' of the occupied area, scaling the radius, R_k, of such an area ($R_k \propto A_k^{\frac{1}{2}}$) to its population. *A priori*, we would expect a strong relationship between population and area, although previous research in urban allometry has suggested that the precise form of the scaling is problematic. For example, Nordbeck (1971) suggests that the dimension, Δ, should be 3 (while the scaling constant, β, should be $\frac{3}{2}$), since population growth takes place in three dimensions. Results from urban density theory also suggest that as cities get bigger their average density increases, but the empirical evidence on this is mixed and is complicated by the definitions of urban area used (Muth 1969). Woldenberg (1973), however, shows quite unequivocally that $\beta \cong 1$, based on an analysis of two large population-area data sets for American cities. Batty and Longley (1994) suggest that the empirical relation between N_k and A_k is of the simplest kind — perfect scaling — with both the theoretical model and empirical evidence suggesting that $\Delta \approx 2$ and $\beta \approx 1$.

The scaling between the urban-field area, U_k, and population size, N_k, is more complicated. As cities grow, their fields become correspondingly larger, growing at a more than proportionate rate. This implies that as cities grow, their field density, $\frac{N_k}{U_k}$, always decreases. For measurement purposes, it is often more appropriate to represent the field area, U_k, in terms of its radius, $R_k = U_k^{\frac{1}{2}}$. Batty and Longley (1994) state the field relationship as:

$$N_k = \varphi R_k^D = \varphi U_k^{\frac{D}{2}} \tag{9.2}$$

where

φ is a constant of proportionality,

D is the scaling constant, and

the fractal dimension lies between 1 and 2.

Further relationships may be postulated between the length of the bounding envelope, E_k, and both the area, A_k, and the field radius, R_k. The envelope defines the outer edge of the cluster. As such, it is likely to be smoother and less circuitous than the perimeter. The bounding envelope is not the perimeter of a settlement cluster, in that any undeveloped interior of the cluster is not detected by the envelope (Figures 9.1a and 9.1b). This suggests that the fractal dimension of the envelope is likely to be smaller than the fractal dimension of a cluster of urban land uses which is juxtaposed with non-urban land uses, but within the enveloping boundary. In the case of the urban area, A_k, we can relate the envelope to the assumed radius, $R_k = A_k^{\frac{1}{2}}$, of the occupied area, giving

$$E_k = \zeta A_k^{\omega} = \zeta A_k^{\frac{\delta}{2}} \tag{9.3}$$

while for the field radius, a similar relation is postulated

$$E_k = \upsilon R_k^{\check{D}} \tag{9.4}$$

where ω, and hence δ in Equation 9.3 and $\check{D}$ in Equation 9.4, can be regarded as 'dimensions' with ζ and υ as constants of proportionality. The log-linearized forms of Equations 9.1–9.4 are given as

$$logN_k = log\gamma + \beta logA_k, \; (\Delta = 2\beta) \tag{9.5}$$

$$logN_k = log\varphi + DlogR \tag{9.6}$$

$$logE_k = log\zeta + \omega logA_k, \; (\delta = 2\omega), \; and \tag{9.7}$$

$$logE_k = log\upsilon + \check{D}logR_k \tag{9.8}$$

Longley *et al.* (1991) and Batty and Longley (1994) describe how the parameters from Equations 9.5–9.8 may be used to measure the size and shape, scale and dimension of urban settlements using vectorized boundary data produced for the UK Department of the Environment (DoE), now the Department of Environment, Transport and the Regions (DETR) (OPCS 1984). A problem with vectorized outlines, such as those shown in Figure 9.1, is that the classification of land use into the 'urban/non-urban' dichotomy is inherently subjective. For example, the UK Department of the Environment definition invokes a number of 'rules of thumb', both to define the extent of the urban areas and to ascribe population figures to the delineated areas. One of these assumptions is that all land which is for practical purposes 'enclosed' by urban land is itself deemed 'irreversibly urban'. This renders the derivation of even crude density measures a suspect exercise, since there are no 'holes' in the urban fabric, and sub-categories of urban land use are not differentiated. The differences between the vector boundary representation and a classified satellite image are illustrated in Figure 9.2. Batty and Kim (1992) and Mesev *et al.* (1995), however, describe two further scaling relations which are applicable to images from urban remote sensing and, through these, identify improved and disaggregated density-size relations. These are of particular use in measuring the densities of different land-use categories, such as those derived in Mesev *et al.* (Chapter 5, this volume).

The first of these measures derives from the urban modelling literature and from the suggestion by Batty and Kim (1992) that the inverse power function

$$p(R) = KR^{-\alpha} \tag{9.9}$$

urban boundary data (DoE)

classified Landsat-TM data   0   kilometres   5

Figure 9.2: Comparison between urban boundary and classified satellite data.

is consistent with fractal geometry conceptions of urban growth, where $p(R)$ denotes the density of occupied space at radius, R. Batty and Kim (1992) show that, given limits on the range of Equation 9.9, the cumulative population $N(R)$ associated with the density $p(R)$ can be modelled as:

$$N(R) = GR^{2-\alpha} \tag{9.10}$$

from which the area, $A(R)$, over which density is defined with respect to distance, R, from the centre is given as:

$$A(R) = ZR^2 \tag{9.11}$$

where a perfect circle of area A would have $Z = \pi$. Further, it is possible to show that the density parameter, α, is related to the fractal dimension, D, such that $D = (2 - \alpha)$. Substituting this into Equation 9.11 gives:

$$N(R) = GR^D \tag{9.12}$$

where

D is the fractal dimension measuring both the extent and rate at which space is filled by population with increasing distance from the centre.

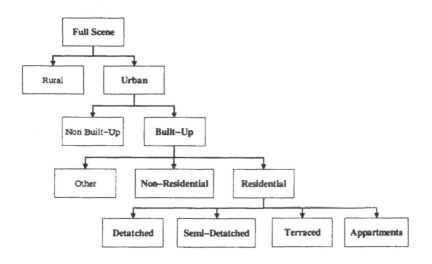

Figure 9.3: Hierarchical segmentation of classified urban categories (**bold** text denotes the categories on which the analysis is carried out).

The empirical dimensions from Equations 9.9 and 9.12 are obtained by regression analysis using the natural log transforms:

$$p_i = lnK - \alpha lnR_i \qquad (9.13)$$

and

$$N_i = lnG - DlnR_i \qquad (9.14)$$

It is these quasi-continuous density measures that will be used in the following section.

9.5 Empirical Case Study: Norwich Revisited

To maintain a degree of comparability with one of the case studies developed in Mesev *et al.* (Chapter 5, this volume), as well as a substantive comparison with previous research (Longley *et al.* 1991), this section will briefly survey the application of the occupancy and density measures (Equations 9.11 and 9.12) to the settlement of Norwich, UK, using the modified maximum-likelihood classification of the 1991 Landsat TM image. In Mesev *et al.* (Chapter 5, this volume) we described how socio-economic information could be incorporated in the image classification process, while Figure 9.3 illustrates how the modified maximum-likelihood classifier can be used to subdivide a full scene into a succession of

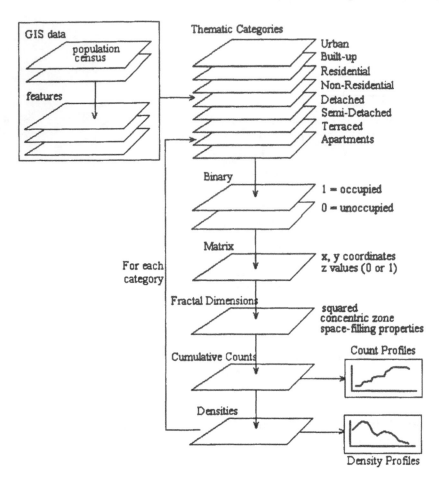

Figure 9.4: Methodology: the conversion of classified thematic categories into fractal dimensions, cumulative counts and density profiles.

urban land-use categories. For purposes of analysis, we have chosen to interpret the outputs of the classification in a deterministic manner; that is for each pixel, the land category for which the predicted probability of membership is highest is taken as the only economic land use in the entire pixel. In practice, of course, a pixel will usually contain more than one residential property, and possibly also a mix of residential and non-residential land uses: however, previous research (Batty and Longley 1986) suggests that a deterministic inter-pretation of such probabilistic forecasts results in patterns of land use which bear closer correspondence with 'real world' configurations. Each land-use category is thus coded as a binary surface, and purpose-written programs are then used to calculate fractal dimensions and to create cumulative count and density profiles. The complete procedure is illustrated in Figure 9.4.

Figure 9.5 shows the 1991 cumulative count profile for the urban and built-up categories. The built-up category may then be decomposed into residential versus non-residential categories (Figure 9.6). Further breakdowns of the residential category (Figure 9.7) identify a relative concentration of apartment properties in the inner areas at the expense of terraced property, otherwise there is a fairly even spatial distribution of all four built forms, as befits a large English market settlement. Corresponding density profiles for the same categories of land use are shown in Figures 9.8 (urban and built up), 9.9 (residential and non-residential) and 9.10 (detached, semi-detached, terraced and apartments). The first of these suggests stability in the density of the 'urban' and 'built-up' categories until $\ln R \approx 5.5$, whereafter there is a quite dramatic tailing off of values. This is consistent with the analogue model results reported by Fotheringham *et al.* (1989) — namely, that urban density gradients within the 'complete' part of an urban settlement remain constant as a settlement grows, and depart significantly from these space-filling 'norms' only in the outer 'growth zones' of settlements in which growth is still occurring. Figure 9.9 suggests that it is the space-filling 'norms' of the residential category that appear to be the cause of such apparent stability. Consequently, there is a need for further research to establish whether space-filling 'norms' characterize any of the constituent non-residential urban land uses (public open space, industrial, transport, etc.). Figure 9.10 depicts considerable instability in the density profiles of the constituent sub-categories of detached, semi-detached, terraced and apartments. Note, however, that the count and density profiles relate to the numbers and densities, respectively, of pixels dominated by each residential category, and not by the imputed raw numbers of properties.

Calculation of the fractal dimensions underlying these profiles can be confounded if observations are used which relate to:

1. central areas where the small number of potential development points are (possibly unrepresentative) high leverage points; and

2. peripheral areas where the growing city is still filling space and development is yet incomplete (see above).

Thus, Equations 9.13 and 9.14 were used only for the scale ranges indicated in Table 9.2, and separate regressions were fitted for the scale ranges separated by ticks marked in these figures. The fractal dimensions for the land uses are shown in Table 9.2. These exhibit a number of characteristics in common with other empirical research into the measurement of morphology, such as the lower fractal dimensions for peripheral zones which, as stated above, are the 'growth zones' of the developing city (Fotheringham *et al.* 1989). Further research is required to ascertain the spatial and temporal transferability of these findings. Other methodological issues raised by this study include the comparability of vector boundary measures based upon Equations 9.5–9.8. Mesev (1995) has begun work in this vein: his results are not reproduced here for reasons of space and lack of direct comparability with the truncated regression results. There is also a need to appraise the process of statistical curve-fitting using regression analysis to profiles such as those shown in Figures 9.5–9.10, and to explore the notion that the fractal dimensions of urban land uses might themselves be a function of scale (Batty and Longley 1994).

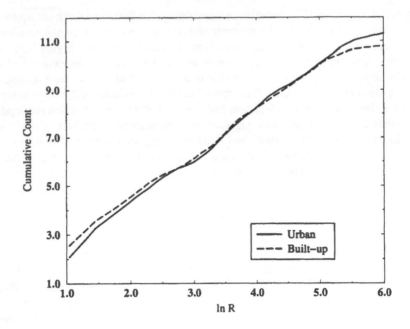

Figure 9.5: Cumulative count profiles of the 'urban' and 'built-up' categories.

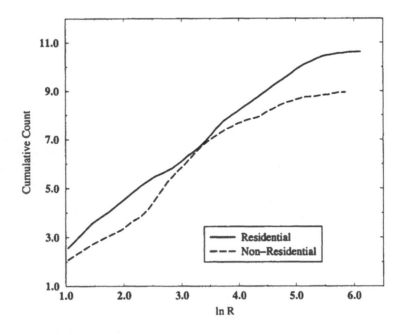

Figure 9.6: Cumulative count profiles of the 'residential' and 'non-residential' categories.

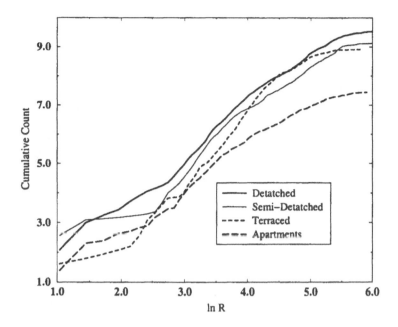

Figure 9.7: Cumulative count profiles of the 'detached', 'semi-detached', 'terrace' and 'apartment' categories.

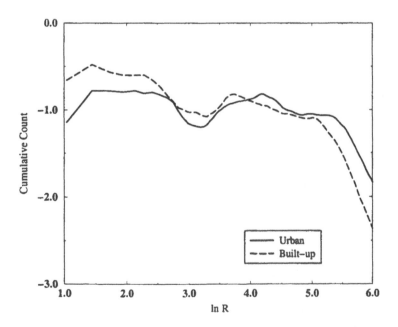

Figure 9.8: Density profiles of the 'urban' and 'built-up' categories.

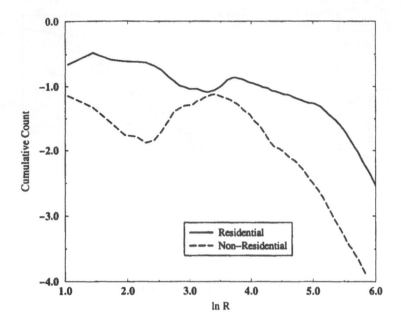

Figure 9.9: Density profiles of the 'residential' and 'non-residential' categories.

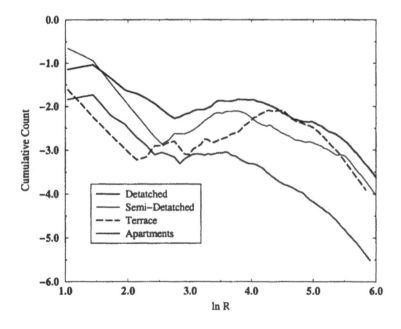

Figure 9.10: Density profiles of the 'detached', 'semi-detached', 'terrace' and 'apartment' categories.

Table 9.2: Dimensions associated with partial population densities

Land Use Category	Count				Density			
	Zone range (R)	ln R	D	r²	Zone range (R)	ln R	D	r²
Urban	2–178	0.693–5.182	1.933	0.977	3–145	1.099–4.977	1.942	0.945
	179–352	5.187–5.864	0.518	0.975	146–358	4.984–5.880	0.646	0.987
Built-up	2–170	0.693–5.136	1.814	0.996	3–108	1.099–4.682	1.867	0.670
	171–319	5.142–5.765	0.262	0.966	146–358	4.984–5.880	0.522	0.984
Residential	2–170	0.693–5.136	1.750	0.955	3–103	1.099–4.635	1.820	0.771
	171–319	5.142–5.765	0.266	0.957	109–319	4.691–5.765	0.542	0.982
Non-Residential	2–55	0.693–4.007	1.936	0.975	2–8	0.693–2.079	1.442	0.955
	56–247	4.025–5.509	0.558	0.911	9–21	2.197–3.044	2.672	0.908
					22–247	3.091–5.509	0.750	0.977
Detached	11–156	2.398–5.050	1.676	0.983	11–38	2.398–3.638	2.356	0.898
	157–316	5.056–5.756	0.335	0.915	39–316	2.664–5.756	1.043	0.929
Semi-detached	9–164	2.197–5.100	1.651	0.917	9–26	2.197–3.258	2.738	0.968
	165–316	5.106–5.756	0.194	0.783	109–319	4.691–5.765	1.116	0.912
Terraced	6–59	1.792–4.077	2.535	0.994	6–63	1.792–4.143	2.535	0.896
	60–102	4.094–4.625	1.231	0.987	64–243	4.159–5.493	0.603	0.967
	103–243	4.635–5.493	0.279	0.834				
Apartments	8–123	2.079–4.812	1.426	0.976	3–25	1.099–3.212	1.527	0.614
	124–262	4.820–5.568	0.398	0.949	8–21	2.079–3.044	0.889	0.971

9.6   Conclusions and Directions for Further Research

The aim of this Chapter has been to demonstrate how, having heeded the tide of relativism, urban remote sensing can contribute towards scale-sensitive measures of urban morphology. Urban remote sensing has hitherto been preoccupied with spectral classifications as the end-product of data modelling, yet there is now a need to develop appropriate spatial analytical methodologies to measure the spatial morphology of classified urban land cover and land use. This is likely to develop alongside the data integration of socio-economic sources, as set out by Mesev *et al.* (Chapter 5, this volume). Fractal analysis is well-suited to the measurement and simulation of fractal forms, and there are enticing prospects for the integration of data-led urban morphology measures with 'top down' fractal simulations of urban structure (Batty and Longley 1994).

Research has yet to develop the important links between spatial analysis and remote sensing, especially in the urban sphere. These links offer the prospect of enhancing our understanding of how urban patterns produced from classified urban land-cover types can describe, and hence prescribe, changes in the morphological and functional characteristics of settlements. There is a need to explore more deeply the ways in which fractal geometry can be used in the spatial analysis of classified urban areas, and how fractal measurement might be used to assess the ways in which patterns of urban land use vary across space and across time.

Our final point is that urban morphology itself is changing, as the monocentric cities of classical urban economics are replaced by 'edge cities' in the US and decentralized, suburbanized morphologies in Europe. It has been argued elsewhere (Batty 1995) that large scale, deductive, system-wide urban planning is a phenomenon of the past, and that detailed, yet generalizable spatial analysis is necessary if planning is to remain a prescriptive force in the planning of 'post-industrial' towns and cities. This research is a timely application of new datasets and analysis methods to a tractable and relevant planning problem: there are also prospects for widening the scale range of analysis to the micro scale (Orford 1999). It is also timely and relevant in the context of the current range of 'sustainable cities' initiatives and, in the spirit of the ESF GISDATA initiative, to tackle the problems of data compatibility across space. Above all, we are beginning to understand that even in the data-rich environment of the 1990s, there are no 'objective', purely technological solutions to the measurement and analysis of urban morphology. Data-led, theory-informed modelling strategies now offer the very real prospect of our developing a classification of urban morphologies based upon a deeper understanding of the ways in which urban processes lead to different size, shape and density relations in urban settlements. If we can understand how cities fill the space available to them, we can begin to reformulate ways in which planning policy interventions might affect issues of spatial efficiency and equity.

9.7   References

Barnsley, M. J., Barr, S. L., Hamid, A., Muller, J.-P., Sadler, G. J., and Shepherd, J. W., 1993, Analytical tools to monitor urban areas, In *Geographical Information Handling — Research and Applications*, edited by P. Mather (Chichester: John Wiley and Sons), pp. 147–184.

Batty, M., 1995, New ways of looking at cities. *Nature*, **377**, 574.

Batty, M., Fotheringham, A. S., and Longley, P., 1989, Diffusion-limited aggregation and the fractal nature of urban growth. *Papers of the Regional Science Association*, **67**, 55–69.

Batty, M., and Kim, K.-S., 1992, Form follows function: reformulation urban population density functions. *Urban studies*, **29**, 1043–1070.

Batty, M., and Longley, P. A., 1986, The fractal simulation of urban structure. *Environment and Planning A*, **18**, 1143–1179.

Batty, M., and Longley, P. A., 1994, *Fractal cities: A geometry of form and function* (London: Academic Press).

Benguigui, L., and Daoud, M., 1991, Is the suburban railway system a fractal? *Geographical Analysis*, **23**, 362–368.

Clarke, G., and Clarke, M., 1995, The development and benefits of customised spatial decision support decisions, In *GIS for Business and Service Planning*, edited by P. A. Longley, and G. Clarke (Cambridge: GeoInformation International), pp. 227–245.

Davis, F. W., Quattrochi, D. A., Ridd, M. K., Lam, N. S., Walsh, S. J., Michaelson, J. C., Franklin, J., Stow, D. A., Johannsen, C. J., and Johnston, C. A., 1991, Environmental analysis using integrated GIS and remotely-sensed data: some research needs and priorities. *Photogrammetric Engineering and Remote Sensing*, **57**, 689–697.

Dutton, G., 1973, Criteria of growth in urban systems. *Ekistics*, **36**, 298–306.

Fotheringham, A. S., Batty, M., and Longley, P. A., 1989, Diffusion-limited aggregation and the fractal nature of urban growth. *Papers of the Regional Science Association*, **67**, 55–69.

Fotheringham, A. S., and Rogerson, P. A., 1993, GIS and spatial analytical problems. *International Journal of Geographical Information Systems*, **7**, 3–19.

Frankhauser, P., 1988, Fractal aspects of urban structures, In *International Symposium des Sonderforchungsbereich 230: Naturliche Konstruktionen-Leichtbau in Architekur und Natur: Teil 1*, Universities of Stuttgart and Tübingen.

Frankhauser, P., 1990, Aspects fractals des structures urbaines. *L'Espace Geographique*, **19**, 45–69.

Frankhauser, P., 1992, Fractal properties of settlement structures, In *First International Seminar on Structural Morphology*, Montpellier, France.

Frankhauser, P., 1993, *La fractalité des structures urbaines*, Ph.D. thesis, UFR de Geographie, Université de Paris I.

Goodchild, M. F., Haining, R., and Wise, S., 1992, Integrating GIS and spatial data analysis: problems and possibilities. *International Journal of Geographical Information Systems*, **6**, 407–423.

Goodchild, M. F., and Mark, D. M., 1987, The fractal nature of geographic phenomena. *Annals of the Association of American Geographers*, **77**, 265–278.

Gould, S. J., 1966, Allometry and size in ontogeny and phylogeny. *Biological Review*, **41**, 587–640.

Hägerstrand, T., 1952, The propagation of innovation waves. *Lund Studies in Geography, Series B: Human Geography*, **4**, 3–19.

Haggett, P., Cliff, A. D., and Frey, A., 1977, *Locational Analysis in Human Geography* (London: Edward Arnold).

Longley, P., Batty, M., and Shepherd, J., 1991, The size, shape and dimension of urban settlements. *Transactions of the Institute of British Geographers*, N.S. **16**, 75–94.

Longley, P., Batty, M., Shepherd, J., and Sadler, G., 1992, Do green belts change the shape of urban areas? A preliminary analysis of the settlement geography of South East England. *Regional Studies*, **26**, 437–452.

Longley, P. A., and Batty, M. (editors), 1996, *Spatial Analysis: modelling in a GIS Environment* (Cambridge: GeoInformation International).

Martin, D. J., 1996, *Geographical Information Systems: Socioeconomic Applications* (London: Routledge).

Mesev, T., 1998, The use of census data in urban image classification. *Photogrammetric Engineering and Remote Sensing*, **64**, 431–438.

Mesev, T., and Longley, P., 1999, The rôle of classified imagery in urban spatial analysis, In *Advances in Remote Sensing and GIS Analysis*, edited by P. Atkinson, and N. Tate, chapter 12 (Chichester: John Wiley and Sons), pp. 185–206.

Mesev, T., Longley, P., Batty, M., and Xie, Y., 1995, Morphology from imagery: detecting and measuring the density of urban land use. *Environment and Planning A*, **27**, 759–780.

Mesev, T. V., 1995, *Urban land use modelling from classified satellite imagery*, Ph.D. thesis, School of Geographical Sciences, University of Bristol.

Mesev, T. V., Longley, P. A., and Batty, M., 1996, RS/GIS and the morphology of urban settlements, In (Longley and Batty 1996), pp. 123–148.

Muth, R., 1969, *Cities and housing: The spatial Pattern of Urban Residential Land Use* (Chicago, Illinois, USA: Chicago University Press).

Nordbeck, S., 1971, Urban allometric growth. *Geografiska Annaler*, **53B**, 54–67.

OPCS, 1984, *Key Statistics for Urban Areas*, Office of Population Census and Surveys (London: HMSO).

Openshaw, S., 1996, Developing GIS-relevant zone based spatial analysis methods, In *Spatial analysis: Modelling in a GIS Environment*, edited by P. Longley, and M. Batty (Cambridge: GeoInformation International).

Orford, S., 1999, *Valuing the built environment: GIS and house price analyis* (Aldershot: Ashgate).

Rickaby, P., 1987, An approach to the assessment of energy efficiency of urban built form, In *Energy and Urban Built Form*, edited by D. Hawkes, J. Owers, P. Rickaby, and P. Steadman (Sevenoaks, U.K.: Butterworth), pp. 43–61.

Thibault, S., and Marchand, A., 1987, Reseaux et topologie, Technical report, Institut National Des Sciences Appliques de Lyon, Villeurbanne, France.

Whyte, L. L., 1968, Introduction, In *Aspects of Form*, edited by L. L. Whyte (London: Lund Humphries), pp. 1–7.

Woldenberg, M. J., 1973, An allometric analysis of urban land use in the United States. *Ekistics*, **36**, 282–290.

CHAPTER 10

Predicting Temporal Patterns in Urban Development from Remote Imagery

Michael Batty and David Howes

10.1   Introduction

In recent years a flood of digital data has become available, as various agencies begin to utilize the power of the Internet in data transfer and exchange, as new forms of data collection devices such as GPS (Global Positioning System) come into routine use, and as agencies and organizations such as tax authorities and realtors begin to develop comprehensive databases and listing services. Integration of diverse data sources at the urban level is opening up dramatic opportunities for urban researchers and planners to develop many new types of application. Combining data sets which have hitherto not been related enables new checks on accuracy and precision, methods for providing missing data in one source from another, and predictions of data outside the immediate domain of interest. At the other extreme, such integration enables new theories of urban form and functioning to be formulated and tested, such as those which have appeared recently in nonlinear dynamics (Dendrinos 1992, Lepetit and Pumain 1993) and urban morphology (Batty and Longley 1994, Frankhauser 1994). In a middle ground between such pragmatic and theoretical concerns, the emergence of new styles of analysis close to the data, which enable explorations of spatial pattern (Fotheringham and Rogerson 1994), seeks to use new methods of making comparisons and combinations of diverse digital data in informed visual ways.

Yet much if not most of these data concern what we actually observe in terms of the physical city; rarely do we see much of its socio-economic functioning through these data, as this requires human survey for specific compilations. In short, digital data such as the UK Ordnance Survey's ADDRESS-POINT™, remotely sensed imagery, and digital elevation models all concern the physical form and stock of the city, not the activities and functions which provide the dynamics within which cities work on a daily basis. These data must be collected in different ways by various forms of human transactions processing, by census. To provide better pictures of the city, these two generic sources of urban data must be put together and there has already been a little progress towards such integration in pragmatic ways (Bracken 1994). But these more ambitious integrations are only just beginning. In this chapter we will focus on integrating purely digital data, in this case data collected at the parcel level with data which are routinely sensed by satellite at a comparable scale. Ultimately, of course, the challenge of linking remotely sensed to data to mainstream urban analysis will be one of linking the physical to the socio-economic but, for the moment, there is the more obvious and pragmatic task of linking remote sensing (RS) data to other digital sources.

There is a third consideration. The 1990s saw a proliferation of concern for treating the dynamics of cities through urban change. As the pace of change quickens everywhere and as communications become instant and global, our quest is to provide a much better handle on urban dynamics and the kinds of equilibrium or otherwise which characterize the city. The detection of urban change using RS data is an obvious strategy and this is a recurrent theme in the other chapters in this book; see also Lo (1995) and Mesev *et al.* (1995). However, in this context, we will seek to develop a somewhat different view of the temporal structure of urban form, exploring the dynamics of the city from a snapshot in time, from data which provide the temporal stratification of the city at a given instant. This is part of a wider project which seeks to use such data to infer the stability of the urban form of the city using recent ideas from weak chaos theory (Bak and Chen 1991, Batty and Howes 1995a). More simply, it is also part of a concerted attempt to enable new ways of visualizing urban dynamics from fixed points in time (Batty and Howes 1995b).

In fact, to anticipate our analysis here, we will show that we can generate excellent classifications of urban development by temporal period from RS data, in this case at least. If our results are generalizable, this might suggest that such classification be the norm for defining urban structure rather than attempting to define land use categories or other such characte izations. Our immediate goals are, however, quite pragmatic: to provide an initial evaluation of the use of RS data to classify development by age; to integrate parcel-based data in such classification; to provide a method for predicting missing temporal data within the data set and outside it; and to develop new ways of visualizing our existing digital data alongside RS data.

We will begin by outlining some of the particular characteristics of the physical form of cities which might be picked up by RS imagery. Next we will note the type and quality of data which we have for Buffalo, emphasizing the way the growth of the city over the last 250 years suggests how we should organize our data into time periods. This provides the context for an initial supervised classification of a 1992 Landsat Multispectral Scanning System (MSS) scene using the maximum-likelihood classifier within ERDAS Imagine™on four temporal classes. The results imply that the first and last two classes should be combined and, for completeness, we then illustrate a further classification of these classes into one which distinguishes the entire urban area from water, cloud and rural areas which serve as our benchmark classes throughout. Finally we show how the predicted four classes can be used to fill in an area of missing data. Our conclusions simply chart the way forward in terms of this project from which this is our first output.

10.2 Temporal Change in Urban Systems

In this context, the way we define temporal change is largely through the physical character-istics of the city. Our immediate perceptions of the typical western city relate to architectural style, the density of building and traffic, perhaps differences in building materials. At the next level down, we might distinguish between different physical forms associated with dif-ferent periods of building, the way the transport systems have developed in terms of typical transit and other communications systems. What is immediately clear is the way in which the city is an assemblage of physical forms from the past. The traces of this old world lie everywhere, and many of them must clearly be detectable using RS imagery. In more sweeping terms, the aggregate form of western cities is clearly a product of individual and institutional wealth and mobility, overlaid of course on culture (see Weber, this volume).

In terms of physical form — and this is very true for the North American city — the early industrial city was dominated by transit systems and a radial focus on the central business district (CBD). In contrast, the forms of the last fifty or more years are dominated by the diffusion of communications, by lower densities of development, by a massive splintering of the central core and the emergence of satellites of commercial development. This is clear in terms of our applications to Buffalo, which is one of the best examples of how an early industrial city has been transformed during the last 50 years by sprawling low density suburbs. Yet the physical form of the early city prior to the 1920s and 1930s still exists with an increasingly sharp distinction between the old and the new.

One very obvious temporal distinction which is reflected in differences in densities, building materials and in transport systems lies astride the period of the 1920s and the 1930s. However further subdivision of this temporal divide is more problematic. There is some logic in dividing the earlier from the later automobile period — the 1960s and be-yond from the earlier period of the 1930s. But this is not that clear for the emergence of the post-industrial city in the form of 'edge' (Garreau 1991) and 'global' cities (Sassen 1991) which only began in earnest in the mid-1970s. The period prior to the automobile, too, is complicated in that transit improved dramatically from the 1860s. In the next section, we will make some decisions as to how fine our temporal grouping should be based on these concerns but also on the growth profile of the Buffalo region. There is also now a sense in which cities might very suddenly change their form as telecommunications begin to affect dramatically the way we work and trade with one another and this might suggest that a recent division from the 1980s onwards might be sensible (Graham and Marvin 1994). Physical form of residential construction has changed quite dramatically during the last decade in that the outer suburbs of the North American city have come to be characterized by very large low density parcels interspersed with pockets of cluster housing which reflect both an ageing and younger, more transient demography. It is possible that RS imagery might increasingly be able to detect such differences, especially if supported by other individual data sources at the parcel level.

In detecting the temporal structure of development, there is a clear correlation with density, and this involves important questions of scale. As the scale at which development is detected and measured is aggregated, the definition of the objects being measured also changes. At the level where individual buildings can be seen, at levels of 10m and below, density which reflects the existence of urban structures is quite different from the parcel level which is associated with a decade of magnitude difference in scale (100m and above). To an extent then, the most appropriate temporal grouping depends on the scale at which the analysis must take place, and arguably, coarser scales (of 100m or more) are better suited to this kind of analysis than the finer (at 10m or so). Moreover, it is across this coarser scale that the biggest differences in residential construction have taken place during the last 100 years (Herman *et al.* 1988). What this discussion suggests is that we should undertake some very detailed analysis of how density and construction have changed through these scales during the last 250 years. This would enable us to determine the most appropriate scale for this analysis and the way in which plot size and the structures and vegetation contained therein are likely to be sensed by the radiance values associated with RS imagery. We have not been able to do this as yet but in more detailed work, using the new generation of very high resolution satellite sensors, this will be important.

Our quest in this chapter is to provide some preliminary evidence that temporal structure is worth classifying. But we also hope that RS imagery might enable us to identify new trends in development which are easy enough to conceptualize, but difficult to determine in that our experience of the detailed physical evidence is limited. To proceed, we need to look at the data available from Buffalo, which combines two sources — parcel data digitized manually by tax assessors and RS imagery from routine Landsat coverages.

10.3   The Buffalo Data

The basic data set which we are using to explore the urban growth dynamics of the Buffalo metropolitan area (population circa 1.2 million in 1990) is based on all properties upon which tax was levied in 1989. These data identify parcels in terms of their centroids and their size (depth and width). They are also classified by various measures of tax payable and land value but for our purposes, the year of construction of the structures which comprise each property are of most use with division by type of land use also being relevant. The data set was purchased from the State Equalization and Assessment Board and comprises some 336,334 taxable properties assessed in 1989. The oldest structure within the database is estimated at 1750 with properties being recorded in every year since 1799. One feature of these data is that although the locations of all the parcels are known, some have not been classified with respect to age. A large block of parcels (some 85,879 properties or 25% of the data) which exist in the West Seneca-Cheektowaga municipalities has not been coded and these are clearly visible in all the images of these data which follow. Part of our interest in comparing these data with that which is classified from RS images is in providing some means of estimating the age of these missing data, and we will indicate this as part of our conclusions.

To divide the parcel data into a manageable number of temporal classes, we need to examine the growth regime of Buffalo which this data set implies. Of course, these data

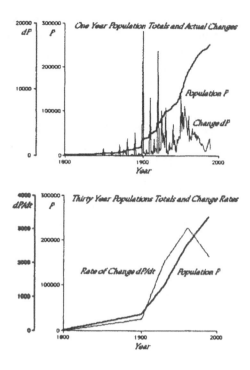

Figure 10.1: Growth profiles of development in the Buffalo Metropolitan region.

do not reveal the actual pattern of growth but simply that development which still remains from the evolution of the metropolitan area over the last 250 years. Much development has been destroyed, built on top of, abandoned or demolished. We argue elsewhere that the temporal structure of development now is likely to provide a good snapshot of how the city has evolved (Batty and Howes 1995a), but this is hardly actual growth. However it is all we have and in any case for these narrower purposes, we are not overly concerned with whether or not we can build good dynamic urban models from these data. In Figure 10.1, we show the growth profile or growth strata associated with these data from which it is clear that Buffalo grew dramatically from the 1880s onwards until the 1960s, whence the growth has been considerably less. Note that in the upper diagram in Figure 10.1, the incremental growth (dP) which is the addition of development in each year, is extremely noisy; the several spiked values indicate that there has been a tendency to overestimate the year of construction of properties at the 10 year period mark, although the trend does show increasing development until the 1950s and from then on a decrease in the rate of this increase. Note also that the data in Figure 10.1 represent the population of properties or parcels, not people.

From a general knowledge of the growth of the urban region, it is clear that the city did not really begin to grow dramatically until the 1860s with the heyday of this growth in the early years of the 20th century. Growth slowed in the 1920s but picked up again in the late 1930s and continued through the Second World War years until industrial decline of the manufacturing base of the city began in the 1960s. Since then the region has begun to lose population. We are also concerned here with aggregating enough development into each temporal class to make the estimation meaningful and a compromise between the classes suggested by the growth profiles in Figure 10.1 and this equal allocation implies the following classes:

Class I:	all development until 1900 — the early industrial city
Class II:	1901 to 1930 — the mature/established industrial city
Class III:	1931 until 1960 — the early automobile city
Class IV:	1961 to 1989 — the mature automobile city

In the lower diagram in Figure 10.1, we show the growth profile (P) associated with these four classes as well as the annualized rate of change ($\frac{dP}{dt}$ or, more properly, $\frac{\Delta P}{\Delta t}$) which confirms the interpretation of the growth regime just described. Note that the number of parcels in each of these four classes is respectively: 35374, 61099, 90936, and 62913 which is sufficient for acceptable estimation.

There is a major problem in visualizing such large data sets which is a familiar one in remote sensing. As the data are based on point locations, aggregation to some grid, which for visual purposes is at the same resolution as the viewing device, changes the meaning of what is being described. We explore this issue in more detail elsewhere (Batty and Howes 1995b) but, as we noted in the last section, there is, in this case, a happy coincidence of the meaning we are seeking in terms of our notion of density (at the parcel level which implies a magnitude of scale at around 100m, not 10, not 1000) and the scale of the RS image we will use (at around 82m resolution but re-sampled at 65m). In fact, in the subsequent discussion we adopt two scales: that of 65m, which is the effective resolution of the image, and that of 100m, which we consider a better representation for the parcel data. This enables some sensitivity analysis to be accomplished, and this is described below. We only show the parcel data at the 65m level in Figure 10.2 which, as in all subsequent screen dumps, is taken as the untransformed grey-scale equivalent of the colour image in which it was displayed. Note the missing data in the rectangular blocks in the central part of this image.

These parcel data are used in our subsequent classification as independent data for the maximum likelihood estimation. This enables the four temporal classes to be extracted from the data in a Landsat MSS scene for the Buffalo region which was recorded on 18th June 1992. The best way to illustrate this image is through its final classification. We have taken the best of all the classifications we have attempted for the four class problem — this is based on using 3 distinct areas for each time period taken from the land parcel data (the so-called training classes) at the 65m level. We refer to this as the image I_65_6. We show this image in Figure 10.3 from which it is clear that besides the four temporal classes of development, we have also classified rural/agricultural use, water, and cloud (and shadow) separately. We use these latter categories as the benchmarks for our urban classes. Note the obscuring interference of cloud cover in the east and south east of the region and the relatively good classification of the water areas which define Lake Erie, the Niagara River, and Grand Island in the north west of the image. Note also the grid road system in the Buffalo Urban Area in the centre north of the image.

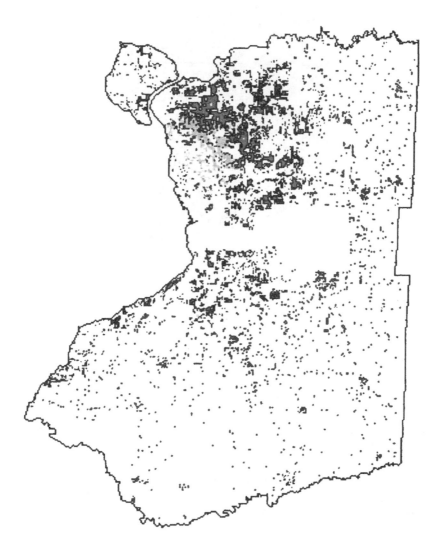

Figure 10.2: Temporal structure of the land parcel data in four age classes. Key: black = 1750–1900; dark grey = 1901–1930; mid-grey = 1931–1960; light grey = 1961–1989.

Figure 10.3: Classification of the MSS image at $65m$ level into four age-classes. Key: black = 1750–1900; dark grey = 1901–1930; mid-grey = 1931–1960; light grey = 1961–1989.

The Landsat MSS data consist of four bands covering wavelengths associated with the electromagnetic spectrum: bands 1 and 2 with wavelengths $0.5\mu m$–$0.6\mu m$ and $0.6\mu m$–$0.7\mu m$ respectively in the area of the visible spectrum, and bands 3 and 4 at $0.7\mu m$–$0.8\mu m$ and $0.8\mu m$–$1.1\mu m$ respectively which are in the near-infrared portion of the spectrum. In general, it appears that bands 1 and 2 are highly correlated, as are bands 3 and 4, but that 1 and 2 are better for discriminating urban materials, whereas 3 and 4 are better for vegetation (Lo 1986). In effect, we have used all four bands in the following classifications. Finally, we should point to some other digital data which we have used variously in this project. We use the TIGER/DLG line files quite routinely as a detailed source of data to rectify images and to provide good visual checks on the geometry of urban form at the most detailed level. These data record features down to the block level (around $500m$ level) and, in fact, in an RS context where road lines and other transport routes are relevant to the sensing, these provide a useful visual check on relative accuracy.

10.4   The Initial Supervised Classification

The process of classification which is used here is based on standard RS approaches which are embodied in most of the available proprietary software. In this context, we used ERDAS Imagine™ running on Sun Microsystems™workstations, and after some experimentation with the data, we decided to classify using all four bands in each image by the maximum-likelihood method. This is a method of supervised classification in which a fixed number of sample regions associated with each potential class of the image are used to train the classifier. These training samples are, in effect, the independent data used in the statistical estimation, and their effect on the quality of the estimation is crucial. The method used is a standard one and a brief overview is given in Lillesand and Kiefer (1994).

In this context, we use the parcel data to identify the training samples for which we know the exact temporal class. Clearly we could use the entire parcel data set for training the classifier and this would give us a very comprehensive method for estimating data outside the parcel data set, but within the image itself. A comparison of Figures 10.2 and 10.3 reveals that the image covers a wider area than the parcel data set except on the southern boundary where the parcel data sets extends beyond the image. However, there is no way within the software to do this routinely and thus, so far, we have simply used the software in the way it was originally intended where the number and size of training samples is much less in terms of extent than the image itself. In the four temporal class problem, we have fixed the number of samples as three for each class, and then six. Concatenating these various assumptions leads to four classification problems which we have coded as follows:

I_65_3: 3 training samples at 65m level
I_65_6: 6 training samples at 65m level
I_100_3: 3 training samples at 100m level
I_100_6: 6 training samples at 100m level

There are many aspects of these classifications that we might examine which involve various statistics pertaining to the pattern evident in the RS data. Here we will show just one.

It is possible to examine the correlation between any two of the four different bands which make up the image and to see how the classification which maximizes likelihood discriminates the various classes portrayed on the scatter plots of radiance values in the two spectral bands for each pixel. In fact, in our analysis, bands 1 and 2 are highly correlated and the four classes are not well discriminated in this space. The discrimination is higher between bands 3 and 4 which are also correlated, but between 1 and 3, 1 and 4, 2 and 3, or 2 and 4, the discrimination is much clearer with cloud, cloud shadow and water occupying very distinct regions of the feature space, and rural to urban falling along a distinct continuum. Taking the probability contours at 3 standard deviations from the mean of each class, we are able to define the classic hyper-ellipse profiles for each class. In our analysis, water, cloud, and cloud shadow are very clearly separate, while rural is separate from the first three temporal urban classes but overlaps the fourth a little. Likewise, the first two and the last two temporal classes overlap but there is little overlap between these two sets. This separability is borne out in all the four classifications as we shall see below.

There are three simple comparisons we can make which show how good these classifications are. First, the number of pixels predicted for each class can be calculated: these are shown for each of the four classifications in the rose diagram which is presented in Figure 10.4a). In general, what this figure shows is that as the number of training samples increases and as the level of resolution of the training sample data is coarsened, the number of pixels predicted for the total of all four urban classes increases quite dramatically. Interestingly, the number of pixels is greater and increases more for the 100m images in terms of Classes

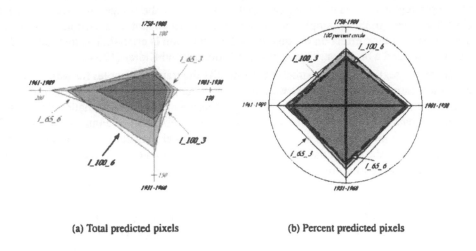

(a) Total predicted pixels   (b) Percent predicted pixels

Figure 10.4: Total predicted pixels and percent predicted pixels in the four-class training sample.

I and III, while the same is true for Classes II and IV in the $65m$ images. Such sensitivity is quite surprising for it suggests that the number and size of training classes is a very important determinant of both the total numbers of pixels predicted for any one class, as well as their relative distribution between classes. The number of pixels in the training classes which are predicted accurately can also be counted and a rose diagram of these percentages is shown in Figure 10.4b). The number predicted varies between about 60% and 80%, but there is a systematic loss of accuracy as the number of training samples is increased. This is certainly due to the greater heterogeneity of the pixel values within the training areas as the number of training areas is increased. However, this degree of variation is perhaps surprising in view of the fact that the training areas are exactly known, and are small in both number and extent.

The last measure of performance involves a more detailed analysis of the extent to which the training samples are misclassified. In Table 10.1, we show matrices which cross-classify these errors in percentage terms. Note that the main diagonals of these matrices form the points on the rose diagram in Figure 10.4b). It is quite clear that the $65m$ images provide better classifications of the training samples than the $100m$, with Class III performing best. For the $100m$ values, Class II is best. However, it is the clear nature of the misclassification which is most revealing. In all four cases, the misclassifications are virtually identical: Class I misclassifies into Class II and II into I while Class III misclassifies into Class IV and vice versa. In all cases, if these portions of these matrices are aggregated, over 85% of all pixels in the training samples are predicted correctly. This is an amazing result.

It suggests that we should combine Classes I and II, and Classes III and IV into two new classes which means that all development to 1930 should be one class, from 1931 to the present the other class. This, of course, is the division between the city based on transit

Table 10.1: Percent misclassification of the training samples for the four urban classes application.

Class	I_65_3				I_65_6			
	I	II	III	IV	I	II	III	IV
I	**70.54**	15.38	0.00	0.00	**73.63**	16.46	0.25	1.00
II	15.59	**83.17**	0.87	0.43	12.53	**79.30**	4.91	0.00
III	2.88	1.44	**93.51**	5.43	1.65	2.01	**80.63**	14.56
IV	0.89	0.00	5.63	**89.13**	2.20	2.20	14.21	**73.18**
Class	I_100_3				I_100_6			
	I	II	III	IV	I	II	III	IV
I	**61.81**	17.97	2.19	1.65	**64.13**	19.01	1.67	3.27
II	19.53	**77.38**	4.76	0.41	18.26	**75.62**	5.00	0.58
III	3.54	2.54	**72.59**	22.50	1.90	2.62	**73.09**	23.04
IV	5.12	2.11	20.33	**71.88**	5.71	2.75	19.86	**69.59**

Key to classes: I = 1750–1900; II = 1901–1930; III = 1931–1960; IV = 1961–1989.

and that based on the automobile, or more loosely between the industrial and emergent post-industrial city. It is remarkable that an RS image should be able to reveal this socio-economic distinction. Our new quest then is to aggregate in the way the analysis suggests and to measure the improvement. Before we continue, however, there are two points to note. First, in all these classifications and those below, we also classify rural areas, water areas, cloud cover, and cloud shadow. In the case of water, clouds and shadow, the predicted performance in terms of the training samples is near perfect and it is well over 90% for the rural areas. Henceforth, therefore, we will graph the predicted number of pixels using cloud and cloud shadow combined with water and rural areas as benchmarks, where appropriate. Finally, for completeness, we show the four classifications in Figure 10.5 which really only gives an impression of density rather than distribution, due to the fact that the grey tones have too little contrast.

10.5 Recombinations and Reclassifications

Classes I and II have been combined into a new Class I′, III and IV into a new Class II′, and the analysis simply reworked in the usual way. In Figure 10.6a), we show the total number of pixels predicted using water, and cloud with cloud shadow (which give near identical numbers between classes), as benchmarks in the rose diagram. The analysis bears out that of Figure 10.4 in that the 65m samples lead to fewer pixels being predicted overall for the more recent (1931–1989) temporal class with slightly more for the earlier class. Overall, the 65m image would appear to predict better the observed values of pixels. We do not, in fact, have the observed number of pixels for these images. All we have is some sense from the parcel data as to their density at this scale. It is from this that we think that in all cases reported here, the RS image classifies more pixels into any of the urban classes than we observe in reality, but the 65m, 3 training sample classification appears the closest.

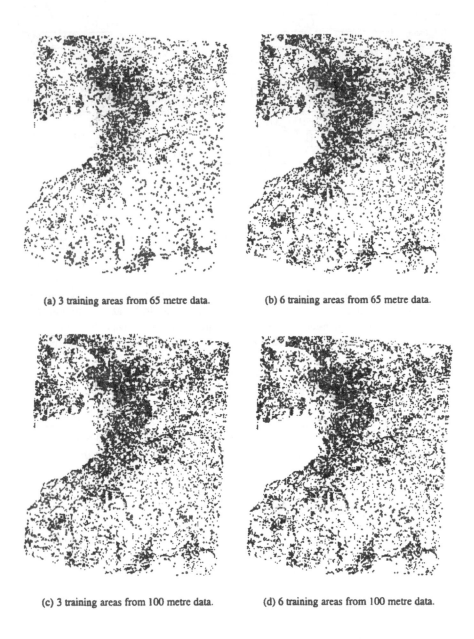

(a) 3 training areas from 65 metre data.

(b) 6 training areas from 65 metre data.

(c) 3 training areas from 100 metre data.

(d) 6 training areas from 100 metre data.

Figure 10.5: Classification into four temporal classes. Key: black =1961–1989; dark grey = 1931–1960; mid-grey = 1901–1930; light grey = 1750–1900.

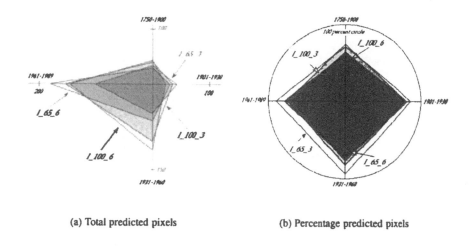

(a) Total predicted pixels   (b) Percentage predicted pixels

Figure 10.6: Total predicted pixels and percent predicted pixels in the two-class training samples.

These results are further confirmed when the percent of pixels misclassified is examined, first in the rose diagram in Figure 10.6b) which is based on the main diagonals of the four misclassification matrices shown in Table 10.2. Table 10.2 shows the quite remarkable fact that hardly any pixels in the training samples for any of the four classifications are misclassified between the two temporal classes. In all cases, over 92% of the pixels in the training samples are correctly classified, with the $65m$, 6 sample classification giving over 99% accuracy for each of the two urban classes. This drops to 96% for the 12 samples, and falls further to 93% and 92% for the respective $100m$ classifications. Note that as classes have been combined, the training samples from the previous four-class problem have been combined, thus leading to 6 from the two classes of 3, and 12 from the two classes of 6. It would be hard to better this type of classification, although we have yet to develop a more rigorous test using the parcel-based data. The classifications are shown in Figure 10.7.

For completeness, but also because considerable effort is put into classifying urban areas from RS imagery, we have combined the two Classes I' and II' into a new urban Class I'' which we now compare against the rural class which we call Class II''. The rose diagram giving the number of pixels predicted is shown in Figure 10.8a) where it is very clear how the increase in resolution ($65m$ to $100m$) and number (12 to 24) of training samples shifts the prediction from urban to rural. However, the rose diagram of percent misclassified in these training samples shown in Figure 10.8b) is near perfect; this is borne out in Table 10.3 where the $65m$ level classification are predicted with 100% and 99% accuracy for the 12 and 24 samples, respectively. This falls to 98% for the $100m$ samples, but in all of these cases the accuracy is extremely good. The urban areas associated with these classifications are shown in Figure 10.9 where, again, it is clear that the smaller and more accurate the training samples, the more compact the urban area predicted and the closer it appears to reality.

Table 10.2: Percent misclassification of the training samples for the two urban classes application.

Class	I_65_6		I_65_12	
	I'	II'	I'	II'
I '	**99.38**	0.57	**96.70**	3.09
II '	0.62	**99.13**	3.30	**95.82**
Class	I_100_6		I_100_12	
	I'	II'	I'	II'
I '	**94.36**	4.78	**94.24**	5.73
II '	5.64	**93.36**	5.76	**92.41**

Key to classes: I' = 1750–1930; II' = 1931–1989.

Table 10.3: Percent misclassification of the training samples for the one urban class application.

Class	I_65_12		I_65_24	
	I''	II''	I''	II''
I ''(Urban)	**100.00**	0.00	**99.53**	0.47
II ''(Rural)	0.00	**99.62**	0.47	**98.92**
Class	I_100_12		I_100_24	
	I''	II''	I''	II''
I ''(Urban)	**98.75**	1.23	**98.90**	1.11
II ''(Rural)	1.25	**98.44**	1.10	**98.47**

Key to classes: I'' = Urban; II'' = Rural.

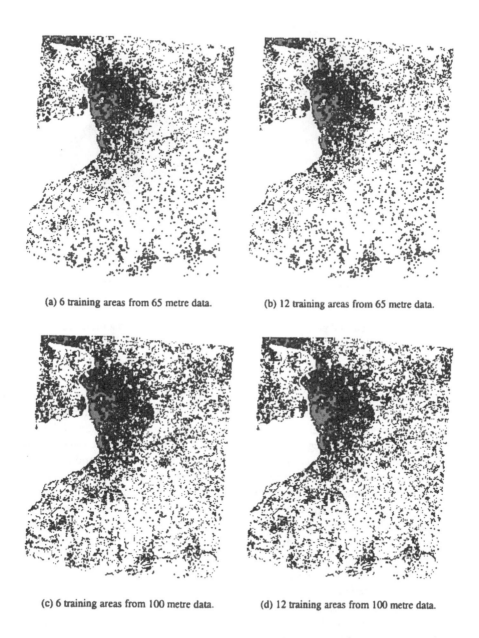

(a) 6 training areas from 65 metre data.

(b) 12 training areas from 65 metre data.

(c) 6 training areas from 100 metre data.

(d) 12 training areas from 100 metre data.

Figure 10.7: Classification into two urban temporal classes. Key: dark grey = 1931–1989; mid-grey = 1750–1930.

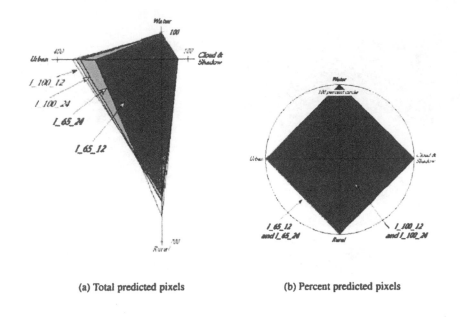

(a) Total predicted pixels   (b) Percent predicted pixels

Figure 10.8: Total predicted pixels and percent predicted pixels in the urban-rural classes training samples.

Finally, we must give an indication of how this type of model might be used in prediction. In our particular applications, there are some very specific problems that can be so addressed. There is the general problem of predicting the age and location of development immediately adjacent to our region of interest. In fact, the parcel data we have is simply for Erie County, which constitutes around 80% of the Buffalo urban region and thus our classifications can help inform the overall pattern of development at a fine scale. Second, Buffalo sits on the border with Canada and the kinds of parcel data we have here are not available there. The images we see, apart from the parcel image in Figure 10.2, all cover areas on the west bank of the Niagara River in the area of Fort Erie and this kind of prediction is perhaps the only way in which we might begin to extend such analysis of urban form across the border. At the same time, however, it can be argued that we require training samples for these areas; as the same independent data set does not exist cross-border, such training would have to be based on other considerations.

Third, as previously noted, there are missing data concerning year of construction in our data set. In Figure 10.10, we have zoomed in on the area where these data are missing and have outlined it on the classified image in Figure 10.3, which is based on I_65_3. What is interesting about this is that the kinds of development which characterize Buffalo — older parcels along main streets, lower densities in more rural areas and so on — are clearly present in the area of missing data. We have also begun to calibrate regression models to estimate this missing data based on variables such as the proximity of residential to

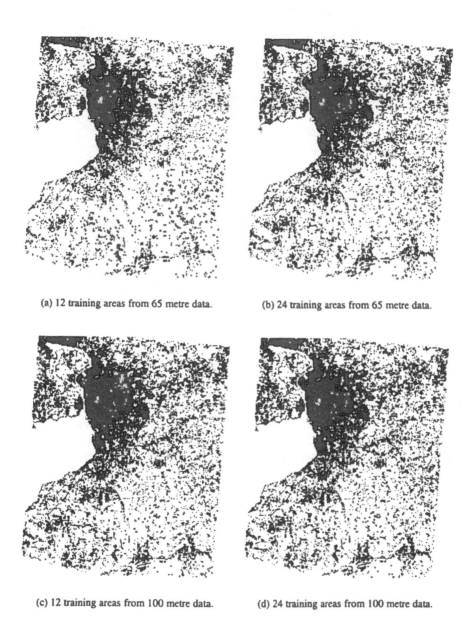

(a) 12 training areas from 65 metre data.

(b) 24 training areas from 65 metre data.

(c) 12 training areas from 100 metre data.

(d) 24 training areas from 100 metre data.

Figure 10.9: Classification into one urban class.

Figure 10.10: Prediction of missing data.

commercial development and distance from the CBD, but so far these are only giving us R^2 values in the order of 30%. Of course, we do not know how good our image classifications are overall. We know that I_65_3 is about 80% accurate in terms of its training samples; thus it is now quite urgent to begin to match the outputs with as much of the parcel-based data so that we can get a much clearer idea as to how good these classifications are.

10.6   Conclusions

This is very much work in progress. We are pursuing this in the spirit that urban remote sensing will generate much better applications for urban analysis if what we know about cities from urban theory is brought to bear on the kinds of objects and areas that need to be detected and classified. One problem which lurks under all this kind of work is the question of scale. Indeed, the very ways in which we are able to display data collected at a much finer resolution than the devices on which we are able to map it poses severe problems of interpretation and definition. There are some very confused notions, for example, concerning the definition of urban density which have been around for 50 years or more, but as soon as detailed data become available such as the RS and parcel-based data used here, these questions will come to the fore.

We have argued the same from the vantage point of new urban theories which deal with urban form (Batty and Longley 1994) and, in an attempt to begin to test theories which are conceived at a much finer scale than those which characterize the mainstream, we have begun to develop morphological analysis of RS images as an alternative to manual digitized data (Mesev *et al.* 1995, Longley and Mesev, this volume). This work, however, has been concerned with how morphologies change over time and is part of the general body of work on urban change and the detection of changing urban boundaries. What we have illustrated here is that we might be able to make as much progress by examining the static properties of cities but from the point of view of their past dynamics. The results we have are encouraging and we will report more as we continue to refine the analysis.

10.7 Acknowledgements

The authors wish to thank Jenny Robinson for her advice concerning image analysis. This paper is based on research carried out at the National Center for Geographic Information and Analysis, supported by a grant from the National Science Foundation (SBR-88-10917).

10.8 References

Bak, P., and Chen, K., 1991, Self-organized criticality *Scientific American*, **264**, 46–53.

Batty, M., and Howes, D., 1995a, Understanding urban dynamics: Exploring spatial evolution through visualization and animation, In *Third Annual Conference on Research into Geographic Information Systems, GISRUK'95*, University of Newcastle upon Tyne, Newcastle, UK.

Batty, M., and Howes, D., 1995b, Visualising urban development, NCGIA, State University of New York at Buffalo, Buffalo, NY. http://www.geog.buffalo.edu/Geo666/batty/buffalo.html.

Batty, M., and Longley, P. A., 1994, *Fractal cities: A geometry of form and function* (London: Academic Press).

Bracken, I., 1994, Towards an improved visualization of socio-economic data, In *Visualization in Geographical Information Systems*, edited by H. M. Hearnshaw, and D. Unwin (Chichester: John Wiley and Sons), pp. 76–84.

Dendrinos, D., 1992, *The dynamics of cities: Ecological determinism, dualism and chaos* (London: Routledge).

Fotheringham, S., and Rogerson, P. (editors), 1994, *Spatial analysis and GIS* (London: Taylor and Francis).

Frankhauser, P., 1994, *La fractalité des structures urbaine*, Collection Ville (Paris, France: Anthropos).

Garreau, J., 1991, *Edge city: Life on the new frontier* (New York: Doubleday).

Graham, S., and Marvin, S., 1994, Telematics and the convergence of urban infrastructure. *Town Planning Review*, **65**, 227–242.

Herman, R., Ardekani, S. A., Govind, S., and Dona, E., 1988, The dynamic characterization of cities, In *Cities and their vital systems: Infrastructure, past, present and future*, edited by J. H. Ausubel, and R. Herman (Washington DC: National Academy Press), pp. 23–70.

Lepetit, B., and Pumain, D. (editors), 1993, *Temporalités urbaine*, Collection Villes (Paris, France: Anthropos).

Lillesand, T. M., and Kiefer, R. W., 1994, *Remote sensing and image interpretation* (New York: John Wiley and Sons).

Lo, C. P., 1986, *Applied remote sensing* (Harlow, Essex, UK: Longman Scientific and Technical).

Lo, C. P., 1995, Automated population and dwelling unit estimation for high resolution satellite imagery: A GIS approach. *International Journal of Remote Sensing*, **16**, 17–34.

Mesev, T., Longley, P., Batty, M., and Xie, Y., 1995, Morphology from imagery: detecting and measuring the density of urban land use. *Environment and Planning A*, **27**, 759–780.

Sassen, S., 1991, *The global city: New York, London and Tokyo* (Princeton, NJ: Princeton University Press).

CHAPTER 11

Modelling Geographical Distributions in Urban Areas

Jean-Paul Donnay and David Unwin

11.1 Introduction

Practitioners of urban geography are seldom remote sensing specialists. Urban modellers and planners, if they use remote sensing at all, employ its by-products rather than the raw radiometric data. These by-products typically fall into two groups:

- those which rely on colour composite images analyzed by visual interpretation, and

- those based on land use/land cover classifications, which are often further processed digitally.

The former are mainly restricted to analogue processes and will not be discussed further. In this chapter we outline some of the ways in which the latter can be used to improve our ability to model geographical distributions in urban areas. Any such use of remotely-sensed data in urban analysis must take into account the nature of these data. The limits to the integration of remotely-sensed data in spatial analysis within a Geographical Information System (GIS) have often been highlighted in the literature (Mace 1991, Ehlers 1995). These result from:

- the geometric characteristics of the image, including the quality of its registration and the compatibility of its resolution with the scale of the phenomenon and the area under investigation;

- the relevance of the attainable land cover/land use classification to the purposes of the study;

- the format of the digital data, which may create difficulties relating to its amount (size) and to its grid (raster) discretization of space; and,

- the measurement scale of the studied attribute. For example, land use/land cover is defined on a qualitative scale for which the number of possible mathematical and statistical operations is restricted.

Problems relating to the geometric characteristics of the images have been addressed in previous chapters, as have various methods for classifying, or improving the classification of, remotely-sensed data of urban areas. What remain to be addressed are the relevance of the nomenclature used, the qualitative scale of the land use/land cover assignments, and their pixel-based discretization of space. Each has far-reaching implications for the subsequent processes of urban modelling and planning.

The relevance of the nomenclature is of prime importance as it affects the legitimacy of any further work on the classified image. The richest urban analyses require a very high discriminatory power, embodying morphological, functional, even legal, criteria which can only be provided by other, exogenous data. Such data are nowadays frequently available in so-called integrated GIS, which combine remotely-sensed images with other spatial information. In an environment such as this, the quality of the classifications used must be assessed in order to prevent the propagation of errors into subsequent processing.

Classified land cover/ land use constitutes qualitative and spatially discrete geographical information. It is easily handled in a raster-based GIS, but such data are of limited direct use to urban planning and modelling. Urban modelling frequently presupposes the existence of quantitative variables that are spatially continuous, while planning makes use of a discretization of space into zones which are very different from those given by image pixels.

In this chapter we demonstrate transformations that can be applied to nominal-scale remotely-sensed data to make them more accessible to urban modelling and planning. First, an attribute transformation from a qualitative land use code to quantitative data can be simply achieved by pixel re-classification using exogenous information. Second, the transformation from a spatially discrete to a spatially continuous distribution can be achieved by a variety of techniques which make use of the concepts of density and potential. After considering the reliability of classified images in the framework of urban planning and modelling, we examine and illustrate these transformations.

11.2   Land Use

In many developed countries, a land use survey constitutes a legal prerequisite to any urban planning process. This is often undertaken by a compilation of maps and aerial photographs complemented by some field survey, but increasingly such land use survey is becoming a typical application of satellite remote sensing. However, this straightforward, if not simple, application of remote sensing meets with two difficulties. The first relates to classification error about which there is a considerable literature in remote sensing (Rosenfeld and Fitzpatrick-Lins 1986, Stehman 1996), although the impact of error in the observations has not, in the past, been an issue addressed within urban planning and its introduction via remote sensing may be unwelcome to urban practitioners. The second difficulty relates to the land use nomenclature used. Instead of there being a single, agreed land use classification, there is a diversity of nomenclatures in urban planning that results from the existence of numerous planning agencies, different administrative purposes, and a variety of other historical and cultural factors. In particular, most of the land use classifications employed in urban planning resort to functional criteria to discriminate among land uses, relying frequently on the four key functions of urbanism, namely dwelling, working, entertaining and moving (Le Corbusier 1957). This functional priority is at odds with remote sensing's basic classifications which are based on the recorded radiometric values. Conversion between the two involves resort to a considerable amount of ancillary data and may lead to higher error probabilities in the final product. Fortunately, urban planning also entails a physical dimension which is unrelated to function but is concerned with the arrangement of the land uses within the urban landscape and which can more easily be addressed using remote sensing.

This distribution of urban land use was subjected to modelling in classical geography, both around (e.g. von Thünen's model) and inside the city where urban structures were deduced from the land use arrangement (e.g. Burgess's model and its successors). Urban models can deal explicitly with land use distribution (Laidlaw 1972), but only rarely do they deal with land use on its own. Generally, they resort to demographic or socio-economic data to enrich the modelling process (Krueckeberg and Silvers 1974, Batty 1976, Longley and Mesev this volume) and it is generally assumed that such complementary data are unproblematic. However, such data, obtained from periodic censuses in the developed world, have a number of disadvantages when used in an integrated GIS (Rhind 1991).

Census data are usually published at a level of spatial aggregation which preserves the confidentiality of the original respondents, but which changes over time and which normally does not coincide with other, equally arbitrary, spatial sub-divisions that may be used for other purposes (Raper *et al.* 1992). The combination of statistical data with land use information derived from remotely-sensed data necessitates the use of a common geography, either by vectorizing groups of neighbouring pixels of the same land use or, more generally, by rasterizing the census enumeration districts. This format transformation not only allows multi-source analyses free from incompatible spatial sub-divisions, but, as shown in the next section, it can also allow a finer distribution of the census derived statistical data across the intra-urban land uses. Finally, provided an appropriate transformation model is known, land use can be used as a surrogate for a great number of other human or physical variables for urban models. This kind of transformation is conditioned by substitution rules and requires appropriate calibration, but it may provide an easy way to change the measurement scale from a qualitative land use/land cover scheme to some quantitative scale on which the substituting variable is coded. Whether it is integrated as a meaningful variable, as a surrogate into the modelling process, or simply used to allow the re-allocation of the statistical values entering into models, it is clear that land use can play a major rôle in urban modelling.

11.3 Densifying Observations using Remote Sensing

A primary use of classifications of land use/land cover derived from satellite remote sensing is to provide data for a large range of urban models by a transformation which changes their measurement scale from nominal categories into attributes lying on binary, ordinal or ratio/interval scales. Such operations retain the basic geography of the distribution but have the capacity to allow urban modelling using much finer spatial resolutions than are commonly employed. To give an indication of the utility of remote sensing in this 'densification' of urban data, in the following sections we review a selection of these processes.

11.3.1   Self-Induced Quantification

The only possible operation applied to qualitative data such as land cover categories is similarity–dissimilarity which, when generalized, can lead to the selection or rejection of a group of items. In turn, this may be translated into a binary scale (1 = selected, 0 = rejected) and so generate a binary image mask. In the raster GIS environment, such masks are the cornerstone of the so-called map- or layer-algebra, allowing Boolean operations to be performed on a pixel-by-pixel basis between two or more maps or layers, and for which specification languages have been proposed (Tomlin 1990). This re-classification process is often the first in a series of operations, but it is worth noting that, because of the constant pixel size, conversion to a binary scale amounts to a quantification of the selected pixels by their pixel area. All that is required to find the area is multiplication by a single constant and thus these binary masks can be used in the computation of scalar density values by use of a quantitative attribute, in a second layer, as numerator for the ratio.

The constant pixel size may also be used to compute the area of the land use polygons made of clusters of neighbouring pixels with similar attributes. This is a two-step operation requiring like-pixels to be gathered into clusters, followed by a simple count of pixels in each group. The allocation of the polygon area to every pixel belonging to it corresponds to a quantification of the attribute which then may be exploited in investigations based on area thresholding or density assessment.

These elementary operations are well known and are not restricted to land use data but, since very often they take place in the early stage of more complex transformations, it is worth reminding the reader of their utility.

11.3.2   Quantification by Substitution

Pixel attribute substitution involves the replacement of a land use/land cover item by a numerical value characteristic of that land use. The replacing variable can be binary, as above, ordinal, or ratio, but in a simple substitution every land use/land cover class is assigned a unique pixel value. This substitution operation is commonly used in accessibility studies, for example, where each land use category is characterized by a penetration or friction co-efficient allowing the determination of a least-cost route across the image (Eastman 1989). Another example is given by inter-visibility problems in landscape analysis. A view-shed is delimited not only by the natural relief but also by the height of the land cover liable to obstruct the lines-of-sight. An estimated mean height is substituted for each land use class and added, pixel-by-pixel, to the digital elevation model in order to provide the required information (Dufourmont *et al.* 1991).

A variation on this simple substitution process is where it is applied only to some of the pixels in the image, because the substituting variable is not significant for all land use categories. For example, it is clear that the three main components of any landscape — i.e. water, vegetation and built-up areas — give complementary, but separate possibilities for substitution. Substitution of the mean biomass for different vegetation items obviously has no meaning for the water or built-up covers. Consequently, substitution often presupposes the identification of alternative transformations which can be performed in different layers separated according to a masking operation of the type mentioned in section 11.3.1.

A second variation on substitution consists of using ancillary information to modify the geographical distribution during the transformation process. A typical example in the field of urban economics consists of the use of the mean land value to substitute for land use categories (see also Brivio *et al.*, Chapter 3 this volume). It is obvious that land use/land cover classes, particularly 'built-up area' and 'building land', cannot be substituted by a single, constant land value over a complete urban agglomeration. A substitution model can be calibrated according to sampled land values from field survey or property files in order to introduce some spatial variation into the mean land values. In the absence of exogenous information, a function of distance from the centre of the city might be used to generate a plausible geographical distribution of the land values.

Apart from a few well-calibrated examples, simple substitution may be seen to be very crude, and indeed it is. Often it is better considered as a transformation from a qualitative to an ordinal scale, rather than to a ratio one.

11.3.3 Reallocation Process and Dasymetric Mapping

The transformation of the measurement scale of pixel attributes in a land use/land cover image can also make use of exogenous census data. However, rather than attempting to alter the land use data into something more useful, the primary aim when combining land use with demographic or socio-economic data is to enable the modification of the geographical distribution of the statistical data. These statistics are available as aggregates over arbitrary census tracts which can raise difficulties in their analysis due to the assumption that the data are evenly distributed within the areal units. An unfortunate consequence of this is that any statistical results are conditional on the areal units employed and thus lack true generality. Space is seen as a discrete phenomenon with statistics tied to distinct areas that can be recorded and stored as polygonal objects in an object or spatial database whose interactions are modelled using discrete approaches. These approaches have the undoubted advantage of fidelity to the original census or survey data, in which all the counts are correct and usually refer to known and mappable areas, but the limitations of the often-used choropleth (area/value) maps are well known. In essence such maps, and any discrete model results which use the same arbitrary geography, may even be random variable-pass filters on the true underlying geography (Baxter 1976, Langford and Unwin 1994).

Except for some applications in the environmental (Eurostat 1993) and agricultural (Tosselli and Meyer-Roux 1990) domains where remote sensing is used to produce or to check statistics, the first role of the land use/land cover map is to give the statistics a geographical base which is more realistic than the original census tracts. A single, striking, example is given by population data, or data for any other socio-demographic index, which logically must relate solely to areas easily discriminated as 'residential' by remote sensing. The number of residential pixels falling into every enumeration district is given by an overlay-like procedure between layers respectively containing the areal units and the remote sensing derived land use. The district population figures can then be reallocated to the residential pixels according to a simple model formalized as follows:

$$p_i = \frac{P_i}{n_i} \qquad (11.1)$$

where P is the population in a given district, p is the population allocated to each residential pixel in a given district, n is the number of residential pixels in a given district, and i is the district index; with $i \in \{1 \ldots I\}$.

A constant apart, and assuming a constant pixel size, this is also an expression for the population density (i.e. $\frac{population}{residential\ area}$) of each district:

$$d_i = \frac{P_i}{n_i.a} \qquad (11.2)$$

where d is the net population density in a given district, and a is the constant pixel area.

Consequently, the masked image derived from the residential pixels holding the population attribute p_i is nothing more than a dasymetric map of the population density. Such maps were originally produced using ancillary information in the form of maps of the residential areas by Wright in a classic paper of 1936 (Wright 1936, Monmonier and Schnell 1984).

This simple reallocation process merely requires a land use/land cover classification of sufficient accuracy to identify residential land within the built-up area. As discussed in other parts of this book, the availability of very high spatial resolution images and the use of sophisticated classification algorithms, together with various types of ancillary data, makes it possible to distinguish between several residential categories (see Mesev et al., Chapter 5, and Barnsley et al., Chapter 7, this volume). In such a case, it may be argued that each residential class should be allocated its own population density value. This leads to a reformulation of the reallocation process, in which a residential class index, r, is introduced in Equation 11.1:

$$p_{r,i} = \frac{P_{r,i}}{n_{r,i}} \qquad (11.3)$$

where r is the residential class index, such that $r \in \{1 \ldots R\}$ and where $P, p, n,$ and i are as given in Equation 11.1.

P_r can be computed given a fixed value of d_r, the population density assigned to the residential class r, knowing the area of this class, A_r:

$$
\begin{aligned}
A_{r,i} &= n_{r,i} \cdot a \\
P_{r,i} &= d_{r,i} \cdot A_{r,i}
\end{aligned}
\qquad (11.4)
$$

with a being the pixel area as above. Thus, the pixel attribute is finally given by:

$$p_{r,i} = d_{r,i} \cdot a \qquad (11.5)$$

The choice of the d_r values is another matter. The problem is constrained by:

$$P_i = \sum_{r=1}^{R} P_{r,i} \qquad (11.6)$$

and generally the number of districts is at least equal to the number of residential classes ($I \geq R$). The logical assumption, from an urban geographic viewpoint, of a common set of d_r values independent of the district subdivision could be met by the solution of the system of I equations with R unknowns:

$$P_i = \sum_{r=1}^{R}(A_{r,i} \cdot d_r) \tag{11.7}$$

Equation 11.5 thus becomes:

$$p_r = d_r \cdot a \tag{11.8}$$

where the reallocation of the pixel attributes is independent of the districts. In practice, however, this approach can generate unrealistic negative density values and does always not perfectly fit the constraint given in Equation 11.6. This is because the strict assumption of common density values is unrealistic and due to errors introduced by an imperfect initial land use/land cover classification. More practically, a set of d_i values dependent on the district subdivision can be adjusted with the weaker assumption of constant ratios, c_r, between the different residential categories. The d_i are then obtained by solving independent equations which respect the total population constraint (Equation 11.6) as:

$$P_i = \sum_{r=1}^{R}(A_{r,i} \cdot c_r \cdot d_i) \tag{11.9}$$

and the pixel attribute is given by:

$$p_{r,i} = c_r \cdot d_i \cdot a \tag{11.10}$$

The coefficient c_r must be specified according to urban standards or density values sampled in the field.

The sophistication introduced into the reallocation process by the discrimination between different residential areas requires optimal conditions in terms of image resolution, classification and ancillary data. In many cases, however, the analysis has to make use of mid- to low-resolution images, poor/inaccurate classifications, or a coarse census-tract subdivision, any or all of which can make the above approach problematic. An alternative is to hypothesize that a relationship exists between population and almost all of the recognized land use categories. For example, Langford *et al.* (1991) employed a simple regression model in which the census tract population was expressed as a function of five land cover categories:

$$P_i = \alpha_0 + \sum_{j=1}^{5}\alpha_j \cdot n_{j,i} \tag{11.11}$$

where α is the regression coefficient, j is the land use index and where P, n and i are as before.

The model gave a good fit to the census-tract population values, with all the coefficients having an intuitively reasonable magnitude and sign. However, unless the ordinary least-squares solution is constrained, the approach can predict impossible negative population densities and is very sensitive to classification error in the pixel counts. A simpler model, used for example in the CORINE Land Cover project (Cornaert 1993), reallocates the district population figures to every land use category according to a plausible scheme of population ratios valid for all districts, such as for example:

Built-up area (all sub-classes included)   80%
Agricultural land covers   15%
Other land covers (but excluding water bodies)   5%

Since the new maps are based on the land use polygons, this method allows visualizations of the population densities which are more detailed than the standard choropleth map of population density per census tract. However, the approach only makes sense when used with large districts comprising all kinds of land cover and is more suitable for regional planning than for strictly urban applications. Finally, it is clear that the reallocation process previously presented (Equations 11.9 and 11.10) could be adapted to deal with all land cover categories instead of just the distinctively residential ones, and in this form would also provide an elegant solution for this scale of analysis.

Whichever of the approaches presented here is adopted, the result is a quantification of the pixel attribute (e.g. a population per pixel) in any zone for which the selected land cover can be described by pixel counts. These numbers have considerable use as a layer in an integrated GIS, for example to make possible operations such as finding how many people live within some arbitrarily area or within a specified distance of some arbitrary point or line. Moreover, dasymetric-like mapping highlights urban morphological structure and is useful for planning, the collection of urban statistics, spatial analysis, simulating the growth of residential areas (Méaille and Wald 1990), the delineation of operational agglomerations (Donnay 1994), and for forcing residential contiguity into grouping and regionalizing methods (Johnston 1970, Weber this volume).

11.4   Surface Transformations

Although the 'substituted' or 'reallocated' pixel values described in Sections 11.3.2 and 11.3.3 can be visualized over a raster, they do not constitute a true surface model in the sense in which that term is usually employed. To produce surface models from these population-per-pixel counts, a further technical step is necessary. In addition, it calls for a re-conceptualization from the discrete model given by the data for the original census tract objects into a continuous surface, or field, model of population density (Muehrcke 1972, Peucker 1972, Rimbert 1990, Couclelis 1992). Surface models, particularly the digital elevation model, have been widely used in physical geography, but more recently they have also been employed as an operational tool in the social sciences (Bracken 1991). As shown below, the availability of urban data in raster form derived from transformations of remotely-sensed images makes the generation of these surfaces both easier and more general.

11.4.1   Surface Modelling from Rasters

Surface generation techniques employ a variety of estimation/interpolation methods which may be classified according to the type of data used, the form of the 'interpolant' or fitting function (Okade *et al.* 1992), and the form of the average weighting used. Moreover, the weighting may be distance- or area-based, while the size of the data subset is a basis for distinction among area-based methods (Watson 1992). Among the main factors which are relevant here, perhaps the most important is the difference between exact and approximate interpolants. Exact interpolants reproduce the original data values, while approximate interpolation may be considered to be a form of data smoothing. Another important distinction must be made between global and local interpolants. The former, classically illustrated by the polynomial trend surface, have their form determined by all of the data values while the latter use, in turn, only the values within a pre-defined neighbourhood of each point.

Considerable use has been made of raster pixels or lattice-point data to represent surfaces. Whether these values are for pixel centroids or grid-line intersections, it is often assumed that they represent point-observations from a continuous spatial surface, and the interpolation involves finding a function which in some sense best represents that entire surface. Nevertheless, to liken a pixel to a point is a gross simplification which usually leads to a preference for an approximate surface-fitting function. Moreover, the objective in this type of work is not to extract a global trend for the surface, but to generate a surface model for whatsoever has been substituted for the pixel land uses. These conditions are best met by low-pass filtering in the image domain.

Image smoothing can be presented as a template technique which, by analogy with time-domain convolution in linear system theory is often called convolution (Richards 1986). Template or window entries, which collectively are referred to as the kernel, are defined and then moved over the image row-by-row and column-by-column. The products of the pixel values covered by the window at a particular position and the template entries are accumulated to produce a new value for the pixel currently at the centre of the template. Consequently, a modified image is produced that smoothes the values according to the size of the template and the specific values that it contains. For a template of dimension $W \times H$ pixels, the new value, s, for the central pixel is given by:

$$s_{r,c} = \sum_h \sum_w p_{h,w} \cdot t_{h,w} \tag{11.12}$$

where s is the new pixel value, p is the original pixel value, t is the template entry, r is the image row index, c is the image column index, h is the template height index; where $h \in \{1 \ldots H\}$, and w is the template width index; where $w \in \{1 \ldots W\}$

With quantitative pixel values, a mean value smoothing uses a uniform template with entries for all h, w:

$$t_{h,w} = \frac{1}{H \cdot W} \tag{11.13}$$

and Equation 11.12 becomes:

$$s_{r,c} = \frac{1}{H \cdot W} \sum_{h=1}^{H} \sum_{w=1}^{W} p_{h,w} \tag{11.14}$$

When the pixel attribute is measured on an ordinal or qualitative scale, an alternative template technique, which is not strictly a convolution operation, may be used in which the pixel at the centre of the template is given the modal class of the pixels covered by the template.

11.4.2 Density Surfaces

In surface models of population, the census data are regarded as an estimate of an underlying, assumed to be correct, distribution of population density. Indeed, if population distribution is seen as a continuous scalar field of population density, the conventional approach, using choroplethic representations and analyses based on the given census tracts, can be seen to be using a set of local integrals on this field in which the limits of the integration (i.e. the boundaries of the areal units used) vary arbitrarily and may even be unknown. To pose the problem in these terms is to illustrate how arbitrary the conventional approach really is and how misleading can be its results. In effect, like can almost never be compared with like.

A partial solution involves the use of regular tessellations, such as grid squares, as the aggregation units. Effectively, this gives the same limits of integration for all of the units and it greatly simplifies the visualization process (Browne and Millington 1983). A grid-cell approach implements what in statistics is referred to as the histogram estimator of the underlying density field (Silverman 1986, Gatrell 1994). This conceptualization of population density as a continuous scalar field in which the quantity mapped formally has the dimensions of an areal density (L^{-2}) has a number of analytical and visual advantages (Martin 1991). Not least of these is the ability to obtain population estimates by integration over any chosen discretization of space, thus allowing population data to be combined easily with other data sets.

The literature presents several attempts to estimate the underlying surface of population density focusing, in particular, on the best way to use the area-based aggregate data as a basic input into this procedure and without making use of any information of the type that might be obtained from remote sensing. An early method, due to Tobler (1979), uses a raster approximation as a step in an iterative procedure which creates a minimum curvature, volume-preserving surface and is called pycnophylactic interpolation. A second method is based on the use of the point centroids of each small census tract. All of the population in each of these is initially assumed to be at this point, but is then spread out from it according to an adaptive distance-decay function and a varying size window. Accumulating these values over a series of grid squares provides a histogram estimate of the underlying density surface (Bracken and Martin 1989, Martin 1989, Martin and Bracken 1991, Bracken 1994a, Bracken 1994b, Longley and Mesev this volume). These two methods are alike in that they attempt to estimate underlying population density surface using as evidence just the original census counts and its basic geography. As has been indicated, both have clear affinities with kernel density estimation in statistics (Silverman 1986).

In order to take advantage of the additional information in the land cover/land use distribution provided by a remotely-sensed image, and of the population reallocation as described in Section 11.3, a smoothing convolution technique can be employed to build the density surface. A kernel of fixed size is moved over the entire map, visiting each pixel in turn to aggregate the population within that kernel, and then estimating the population density for the centre of the pixel as this total divided by the kernel area. This is a direct application of equation 11.14, with the window size expressed in meters:

$$d_{r,c} = \frac{1}{H \cdot W \cdot a} \sum_{h=1}^{H} \sum_{w=1}^{W} p_{h,w} \tag{11.15}$$

where d is the population density, a is the pixel area, p is the population allocated to any given pixel, and where r, c, H, W, h and w are as before.

This technique was suggested in the 1950's when it was called the 'floating grid' technique (Schmid and MacCannell 1955, Porter 1957). The procedure could be improved by adopting one or other of the kernel estimators which ensure continuity in the resulting surface (Silverman 1986). Although technically a surface resulting from an application of this kind of estimator is not continuous, at the scale at which it is visualized this is almost an irrelevance: the critical control on the resulting density surface is not the form of the estimator used, but the width of the kernel over which it is applied, with large kernels giving a greater degree of surface smoothing.

11.4.3 Potential Surfaces

The physical concept of potential was originally introduced into the social sciences as the "population potential" by Stewart (1941). Many reviews focusing on the operationalization and interpretation of his index can be found in both the geographical and the economics literature (Rich 1980, Pooler 1987). The potential has been used for about 50 years as a measure of aggregate accessibility or as a general measure of relative position or location. A scalar field of potential may be further transformed into its equivalent vector field and the vectors interpreted in terms of gradients and flows. It can also be regarded as a distance-weighted smoothed or filtered version of the density surface of dimension L^{-1}or L^{-2} according to the distance power used. The general expression for the potential is given by Equation 11.16 where the first term, expressing the self-potential, has an indeterminacy that must be resolved:

$$pot_i = \frac{w_i}{d_{i,i}} + \sum_{\substack{j=1 \\ j \neq i}}^{N} \frac{w_j}{d_{i,j}} \tag{11.16}$$

where *pot* is the potential value, w is the empirical point value, d is the distance, and i, j is the point index in the data set, such that i and $j \in \{1 \ldots N\}$.

The generation of a population potential field based on a lattice of valued points was first applied at the country level (Nadasdi 1971) and then transposed in raster format to the regional and agglomeration levels (Nadasdi *et al.* 1991). The technique was also rediscovered as a 'new' cartographic method applied to coarse grid cells at the country level by the French statistical agency (Laurent and Tardif 1993). Surface continuity is visually assured by the large distances involved in the computations, but the surface form shows peaks and steep slopes in the vicinity of population poles which, in turn, can be used to rank central places at regional or country scale (Donnay 1995). Although almost all work to date has examined the potential of population, the method allows for potential surface generation based on any quantitative pixel value, and this greatly increases the number of possible applications in urban planning and urban environmental studies (Nadasdi *et al.* 1991, Donnay 1992, Donnay and Nadasdi 1992, Donnay and Thomsin 1994).

The application of the potential to raster data requires a particular operationalization. It is virtually impossible to compute the contribution of all of the pixels in an image to the potential of every pixel. Because the potential can be considered as a smoothing filter, its operationalization in image processing can also be seen as a form of convolution. All that we need to do is to fix the parameters of the function, particularly the window size and the self-potential term. The contribution to the potential only comes down to zero beyond a distance which depends on the range of the variable submitted to the function. However, this distance is always very large, tens of times the pixel size. For the self-potential term, an electrical analogy is often used to avoid a denominator equal to zero. This suggests introducing a distance equal to the half-ray of the circle of same area as the object on which the potential is computed. In this application, the object is a pixel of known and constant area, so the analogy is easily implemented. These considerations lead to a potential function for a raster as:

$$pot_{r,c} = \frac{p_{r,c}}{\frac{1}{2}\sqrt{\frac{a}{\pi}}} + \sum_{\substack{h=1 \\ h \neq r}}^{H} \sum_{\substack{w=1 \\ w \neq c}}^{W} \frac{p_{h,w}}{d_{(r,c),(h,w)}} \tag{11.17}$$

where *pot* is the potential value attributed to the pixel at the centre of the template, *p* is the population allocated to residential pixels, *a* is the pixel constant area, *d* is the distance between pixels, *r, c* are the indices of the central pixel in the template (refer to image row and column indices), *w* is the template width index; with $w \in \{1 \ldots W\}$, and *h* is the template height index; with $h \in \{1 \ldots H\}$.

11.5   Example Applications

11.5.1   Creating a Surface of Population Density

Figure 11.1 shows some results from a dasymetric mapping and surface transformation of the 1991 population census of Northern Leicestershire, UK (Langford and Unwin 1994). The original census data used were at a very coarse level of aggregation (the ward), and these observations were first densified using the techniques outlined in this chapter. First, a Landsat Thematic Mapper image was registered onto the same geographical coordinates as the census data and then a standard principal components analysis (PCA) transformation of these data was used as input to a supervised maximum likelihood classifier to identify pixels that could be classed as residential. Second, using Equation 11.1 the population figures were allocated to the residential pixels falling within each ward to produce a crude dasymetric map of the population-per-pixel. It should be stressed that although it is not easy to visualize this population-per-pixel 'map' is useful in its own right as a layer in further GIS based operations and is a far better representation of the population over this area than is the standard choropleth. Finally, a surface transformation according to Equation 11.15 was applied using kernel areas of 0.25 km² and 1.0 km² and the results visualized in a variety of ways using the UNIRAS™ graphics package. The displays shown in Figure 11.1 uses a fish-net representation of the computed density surface onto which has been superimposed a shade sequence according to contours of the density.

11.5.2 Combining Potential Surfaces

Figure 11.2 shows potential surfaces of built-up intensity and ecological intensity, respectively (Donnay and Nadasdi 1992). They are processed from a classified high resolution SPOT-HRV multispectral plus panchromatic image (re-sampled to 10m resolution) covering the urban agglomeration of Maastricht, in the Netherlands. A grey-scale slicing into five levels is provided for printing purposes only, but the original images were generated with continuous values. In order to compute the potential of built-up intensity, built-up pixels are weighted by three values of the 'floor-area index'. Similarly, pixels classified as vegetation and water bodies are weighted by several levels of ecological interest (ordinal scale) to generate the potential of ecological intensity.

Broadly speaking, the built-up intensity can be seen to be the inverse image of the ecological intensity, so the combination of the two potential images shows limited discrimination in the centre of the town and in the countryside. In the suburbs, on the other hand, the environment is submitted to a strong pressure of the built-up area. This competition is illustrated by a cross-classification of the two potential surfaces (Figure 11.3). Two classes of pressure (high and medium) have been highlighted, as the very high pressure in the city centre and the very low pressure in the countryside are meaningless in this context.

Other combinations of potential surfaces have been performed in urban applications. Some merge land cover potentials to highlight urban structures, while others look for correlations between selected potential surfaces and further surface phenomena such as the urban heat island (Cornélis *et al.* 1998). At the regional scale too, the combination of potentials generated from low spatial resolution classified images has proved useful to cluster patterns of land cover and to identify landscape units and biophysical regions (Binard *et al.* 1997).

11.6 Conclusions

It can be seen that, although they apparently remove the uncertainties related to the modifiable areal unit problem (Openshaw 1984), and facilitate the integration of census data with other kinds of information, approaches which resort to satellite remote sensing substitute a whole set of other uncertainties related to the methods used. Clearly some of the methods we have introduced provide new and original ways to examine urban geographies, but as yet there is little accumulated experience in their use that might enable a balanced assessment of their utility to be made. There have, however, been a few studies which have attempted to assess the relative efficiency of various methods used to estimate population from a classified image. Langford *et al.* (1993), for example, show that the dasymetric method of estimating population density using remote sensing data will almost always improve on the simple areal weighting approach used in most GIS. The impact of differing levels of classification error on the accuracy of the population estimates obtained has been also investigated

218

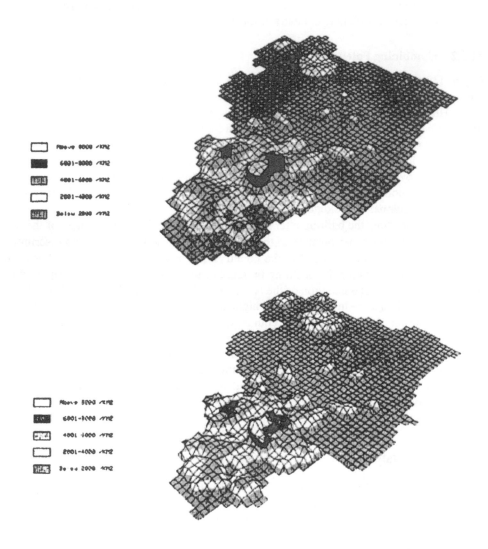

Figure 11.1: Dasymetric mapping and surface transformation of the 1991 population census of Northern Leicester (UK). Top: 1.0 km² window. Bottom: 0.25 km² window.

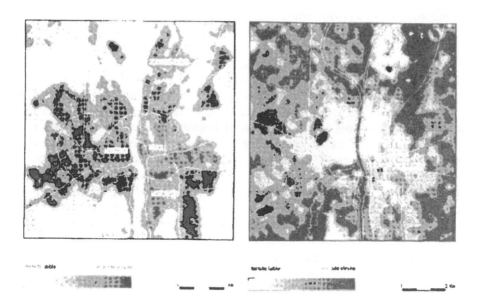

Figure 11.2: Potential surfaces generated from land use data derived from a remotely-sensed image (Maastricht, Netherlands). Left: built-up intensity, Right: ecological intensity.

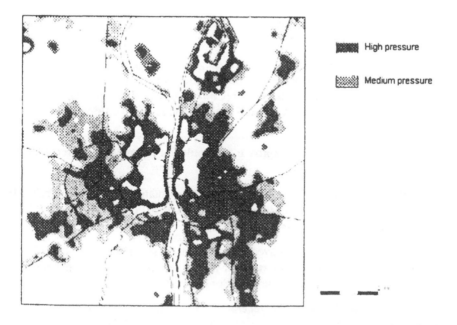

Figure 11.3: Cross-tabulation of built-up and ecological potentials showing competition between the land uses in the suburbs of the town.

by Fisher and Langford (1996), who demonstrate a surprising relative lack of sensitivity of the dasymetric method. Even though individual pixels may have a weak probability of being correctly assigned to a land use category, when aggregated into a target zone for density estimation, the relative frequencies within these target zones do not degrade substantially.

Dasymetric population-per-pixel estimates have utility in many GIS operations, such as polygon overlay and distance buffering, which involve areal aggregations of these pixel values. Estimates of the underlying surfaces of population density or potential using the population-per-pixel values and an appropriate transformation function take advantage of the increased stability of such regional quantities. These can be used to produce intuitively reasonable values which can be displayed in a variety of ways and compared with other urban indicators, such as accessibility and land values. Finally, the additional adaptability of the potential surface model allows it to cope with ordinal as well as quantitative attributes, dasymetric population or other census data, urban indices, or even physical data. This provides a robust exploratory tool for planning and environmental analysis in the urban milieu. Together, the dasymetric method, spatial density estimation and potential field modelling have the added advantage that they require no special software — they can be implemented in almost any raster GIS using standard functions.

11.7 References

Batty, M., 1976, *Urban Modelling: Algorithms, Calibrations, Predictions* (Cambridge: Cambridge University Press).

Baxter, R. S., 1976, Some methodological issues on computer drawn maps. *The Cartographic Journal*, 13, 145–155.

Binard, M., Nadasdi, I., Lambinon, M., Marchal, D., and Donnay, J.-P., 1997, Études multi-échelles: applications aux agglomérations du nord-ouest de l'Europe, In *Télédétection des Milieux Urbains et Périurbains (Collection Actualité Scientifique)*, edited by J.-M. Dubois, J.-P. Donnay, A. Ozer, F. Boivin, and A. Lavoie (Montréal: AUPELF-UREF), pp. 15–23.

Bracken, I., 1991, A surface model approach of population for public resource allocation. *Mapping Awareness*, 5, 35–38.

Bracken, I., 1994a, A surface model approach to the representation of population-related social indicators, In *Spatial Analysis and GIS*, edited by S. Fotheringham, and P. Rogerson (London: Taylor and Francis), pp. 247–259.

Bracken, I., 1994b, Towards an improved visualization of socio-economic data, In *Visualization in Geographical Information Systems*, edited by H. M. Hearnshaw, and D. Unwin (Chichester: John Wiley and Sons), pp. 76–84.

Bracken, I., and Martin, D., 1989, The generation of spatial population distributions from census centroid data. *Environment and Planning A*, 21, 537–543.

Browne, T. J., and Millington, A. C., 1983, An evaluation of the use of grid squares in computerised chloropleth maps. *The Cartographic journal*, 20, 71–75.

Cornaert, M., 1993, Quelques considérations sur les besions d'information pour les politiques d'environnement, ainsi que sur les possibilités de la télédéction, des systèmes d'information géographiques et des systèmes statistiques, In (Eurostat 1993), pp. 81–91.

Cornélis, B., Binard, M., and Nadasdi, I., 1998, Potentiels urbains et îlots de chaleur. *Publications de l'Association Internationale de Climatologie*, **10**, 223–229.

Couclelis, H., 1992, People manipulate objects (but cultivate fields): Beyond the raster-vector debate, In *Theories and Methods of Spatio-Temporal Reasoning in Geographic Space*, edited by A. U. Frank, I. Campari, and U. Fromentini (Berlin: Springer-Verlag), pp. 65–77.

Donnay, J.-P., 1992, Remotely sensed data contributes to GIS socioeconomic analysis. *GIS Europe*, **1**, 38–41.

Donnay, J.-P., 1994, Agglomérations morphologiques et fonctionnelles, l'apport de la télédétection urbaine. *Acta Geographica Lovaniensia*, **34**, 191–199.

Donnay, J.-P., 1995, Délimitation de l'hinterland des agglomérations urbaines au départ d'une image de télédétection. *Revue Belge de Géographie*, **59**, 325–331.

Donnay, J.-P., and Nadasdi, I., 1992, Usage des données satellitaires urbaines de haute résolution en modélisation urbaine: application à l'agglomération de Maastricht. *Acta Geographica Lovaniensia*, **33**, 159–169.

Donnay, J.-P., and Thomsin, L., 1994, Urban remote sensing and statistics: prospective research and applications, In *Proceedings of the Symposium: New Tools for Spatial Analysis*, edited by M. Painlo, Eurostat (Luxembourg: Office for Official Publications of the European Communities), pp. 137–145.

Dufourmont, H., Gulinck, H., and Wouters, P., 1991, Relief dependent landscape typology derived from SPOT data, In *Proceedings of EGIS'91*, EGIS Foundation (Amsterdam: EGIS Foundation), pp. 286–297.

Eastman, J. R., 1989, Pushbroom algorithms for calculating distances in raster grids, In *Proceedings of Autocarto 9*, Autocarto (Autocarto), pp. 288–297.

Ehlers, M., 1995, The promise of remote sensing for land cover monitoring and modelling, In *Proceedings of the Joint European Conference on Geographical Information* (Basel: AKM Messel AG), pp. 426–432.

Eurostat, 1993, *The Impact of Remote Sensing on the European Statistical System — Proceedings of the Seminar, Bad Neuenahr* (Luxembourg: Office for Official Publications of the European Communities).

Fisher, P., and Langford, M., 1996, Modeling sensitivity to accuracy in classified imagery: a study of areal interpolation by dasymetric mapping. *Professional Geographer*, **48**, 299–309.

Gatrell, A. C., 1994, Density estimation and the visualization of point patterns, In *Visualization in Geographical Information Systems*, edited by H. M. Hearnshaw, and D. J. Unwin (Chichester: John Wiley and Sons), pp. 65–75.

Johnston, R. J., 1970, Grouping and regionalising: some methodological and technical observations. *Economic Geography*, 46, 293–305.

Krueckeberg, D. A., and Silvers, A. L., 1974, *Urban Planning Analysis: Methods and Models* (New York: John Wiley and Sons).

Laidlaw, C. D., 1972, *Linear Programming for Urban Development Plan Evaluation* (New York: Praeger Publishers).

Langford, M., Fisher, P. F., and Troughear, D., 1993, Comparative accuracy measurements of the cross areal interpolation of population, In *Proceedings of EGIS'93*, EGIS Foundation (Amsterdam: EGIS Foundation), pp. 663–674.

Langford, M., Maguire, D. J., and Unwin, D. J., 1991, The areal interpolation problem: estimating population using remote sensing in a GIS framework, In *Handling Geographical Information: Methodology and Potential Applications*, edited by I. Masser, and M. Blakemore (Harlow: Longman), pp. 55–77.

Langford, M., and Unwin, D. J., 1994, Generating and mapping population density surfaces with a geographical information system. *The Cartographic Journal*, 31, 21–26.

Laurent, L., and Tardif, L., 1993, Depuis trente ans: dynamiques de l'espace français. *INSEE Première*, 280, 1–4.

Le Corbusier, 1957, *La Charte d'Athènes*, second edition (Paris: Editions de Minuit).

Mace, T. H., 1991, Special issue: Integration of remote sensing and GIS. *Photogrammetric Engineering and Remote Sensing*, 57.

Martin, D. J., 1989, Mapping population data from zone centroid locations. *Transactions of the Institute of British Geographers*, 14, 90–97.

Martin, D. J., 1991, *Geographical Information Systems and the Socioeconomic Applications* (London: Routledge).

Martin, D. J., and Bracken, I., 1991, Techniques for modelling population-related raster databases. *Environment and Planning A*, 23, 1069–1075.

Méaille, R., and Wald, L., 1990, Using geographical information systems and satellite imagery within a numerical simulation of regional urban growth. *International Journal of Geographical Information Systems*, 4, 445–456.

Monmonier, M., and Schnell, G., 1984, Land use and land cover data and the mapping of population density. *International Yearbook of Cartography*, 24, 115–121.

Muehrcke, P., 1972, Thematic cartography. *Resource Paper of the Association of American Geographers*, 19.

Nadasdi, I., 1971, Surface de potentiel de la population de la province de Liège: exemple d'application de la transformation des champs géographiques discrets en champs continus. *Bulletin de la Société Géographique de Liège*, **7**, 51–60.

Nadasdi, I., Binard, M., and Donnay, J.-P., 1991, Transcription des usages du sol par le modèle de potentiel. *Mappemonde*, **3**, 27–31.

Okade, A., Boots, B., and Sugihara, K., 1992, *Spatial Tessellations: Concepts and Application of Voronoi Diagrams* (Chichester: John Wiley and Sons).

Openshaw, S., 1984, *The Modifiable Areal Unit Problem*, volume 38 of *Concepts and Techniques in Modern Geography* (Norwich: GeoBooks).

Peucker, T. K., 1972, Computer cartography. *Resource Paper of the Association of American Geographers*, **17**.

Pooler, J., 1987, Measuring geographical accessibility: a review of current approaches and problems in the use of population potential. *Geoforum*, **18**, 269–287.

Porter, P. W., 1957, Putting the isopleth in its place. *The Minnesota Academy of Science Proceedings*, **25–26**, 372–379.

Raper, J. F., Rhind, D. W., and Shepherd, J. W., 1992, *Postcodes: the New Geography* (Harlow: Longman).

Rhind, D. W., 1991, Counting the people: the role of GIS, In *Geographical Information Systems: Principles and Applications*, edited by D. J. Maguire, M. F. Goodchild, and D. W. Rhind (Harlow: Longman), pp. 127–137.

Rich, D. C., 1980, *Potential Models in Human Geography*, volume 26 of *Concepts and Techniques in Modern Geography* (Norwich: GeoBooks).

Richards, J. A., 1986, *Remote Sensing Digital Image Analysis* (New York: Springer-Verlag).

Rimbert, S., 1990, *Carto-graphies* (Paris: Hermès).

Rosenfeld, G. H., and Fitzpatrick-Lins, K., 1986, A coefficient of agreement as a measure of thematic classification accuracy. *Photogrammetric Engineering and Remote Sensing*, **52**, 223–227.

Schmid, C. F., and MacCannell, E. H., 1955, Basic problems, techniques and theory of isopleth mapping. *Journal of the American Statistical Association*, **50**, 210–2239.

Silverman, B. W., 1986, *Density estimation for statistics and data analysis* (London: Chapman and Hall).

Stehman, S. V., 1996, Estimating the Kappa coefficient and its variance under stratified random sampling. *Photogrammetric Engineering and Remote Sensing*, **62**, 401–407.

Stewart, J. Q., 1941, An inverse distance variation for certain social influences. *Science*, **93**, 89–90.

Tobler, W. R., 1979, Smooth pycnophylactic interpolation for geographical regions. *Journal of the American Statistical Association*, **74**, 519–530.

Tomlin, C. D., 1990, *Geographic Information Systems and Cartographic Modelling* (Englewood Cliffs, New Jersey: Prentice Hall).

Tosselli, F., and Meyer-Roux, J. (editors), 1990, *The Application of Remote Sensing to Agricultural Statistics* (Luxembourg: Office for Official Publications of the European Commnunities).

Watson, D. F., 1992, *Contouring: A Guide to the Analysis and Display of Spatial Data* (Oxford: Pergamon – Elsevier Science).

Wright, J. K., 1936, A method of mapping densities of population with Cape Cod as an example. *Geographical Review*, **26**, 103–110.

CHAPTER 12

Geographical Analysis of the Population of Fast-Growing Cities in the Third World

Yves Baudot

12.1 Introduction

The world's human population is growing rapidly: currently estimated to be about 5.5 billion, it is projected to reach 10 billion by the year 2050. Its geographical distribution is also changing, with the fastest rates of growth occurring in less-developed countries. In Africa, for example, the total population increased five-fold since 1800. This is expected to rise to thirty-fold during this century. By contrast, while developed countries presently account for 25% of world's population, this is likely to decline to 15% by the early part of this century. Much of the projected growth in population will be centred on the world's cities — especially large cities (i.e. those with a population of greater than one million inhabitants). In this respect, the urban population, currently estimated to be about 2.3 billion people, is expected to double by 2020. Perhaps more significantly, it is predicted that 93% of this increase will be associated with Third World cities. If this is the case, approximately two billion people will be added to these already congested urban areas during the lifetime of just a single generation. Not surprisingly, therefore, a recent report by the World Bank concluded that urban growth in developing countries is one of the most explosive problems for the beginning of the next century.

Effective management of the urban population problem demands good diagnostic tools. Accurate and reliable information is also required to quantify the current situation and to predict future trends: information on patterns of land use is one obvious example, while basic data on population (including its spatial distribution and rates of growth) is another. The two are, of course, interlinked since population growth creates pressure on the land in terms of increased demand for new settlements, supporting infrastructure (e.g. the supply of water, food and electricity, and the provision of transportation and waste disposal facilities) and employment. Unfortunately, conventional sources of information on both land use and population are frequently inadequate (Devas and Rakodi 1993). For Third World countries in particular, the necessary data are often outdated, unreliable or, in some cases, simply unavailable. Remote sensing techniques offer an important, alternative source of data in this context. Since several of the other contributions to this volume have focused on deriving information on urban land use from remotely-sensed data, this Chapter will consider the use of remote sensing to analyze the urban population. Before doing so, however, a number of considerations specific to an analysis of cities in less-developed countries will be examined.

12.2   Information Related to Land Use

12.2.1   The Concept of Land Use

The term 'land use' is central to much of the discussion that follows, yet it has a somewhat different meaning for urban planners and for remote sensing specialists. To urban managers, for instance, the term 'land use' incorporates information on the function (economic, social, etc.), administrative status, and ownership of a land parcel. Remote sensing specialists, on the other hand, frequently employ the term quasi-interchangeably with the somewhat narrower notion of 'land cover'. Since the data gathered by remote sensing devices cannot provide all of the information used by urban planners to define and characterize land use, it follows that satellite remote sensing should not be used as the sole source of information in attempts to map land use.

12.2.2   Spectral Information

The sensors mounted on-board Earth-orbiting satellites record data about the spectral reflectance properties of the Earth surface; that is, they measure the amount of surface-leaving radiation in different parts of the electromagnetic spectrum. In general, the information that can be derived from these data relates to the *surface* properties of the features being observed. For instance, if the object is a house, the detected spectral reflectance is primarily controlled by the nature of the materials used to construct the roof, as well as its physical condition and geometric structure (e.g. flat or pitched roof). Urban managers, on the other hand, may require information on type of the building, its date of construction, and number of storeys. Despite the inherent limitations of remotely-sensed data in this respect, several broad categories of land use can be determined from measurements of spectral reflectance. In particular, green vegetation within the urban fabric (i.e. urban open space) is easily identified using data recorded at near infra-red wavelengths. The spectral contrast between vegetation and man-made structures, such as buildings and roads, in this part of the spectrum provides valuable information on the patterns of land cover that can, in turn, be used to distinguish between and to infer different categories of land use.

12.2.3 Spatial Resolution

The spatial resolution of current satellite sensors is perhaps the limiting factor for urban analysis. When this study was carried out, data from the new generation of very high spatial resolution satellite sensors were not widely available, so that SPOT-HRV Panchromatic (P) data (10m spatial resolution) were the finest data available. Since SPOT-HRV P data are, by definition monospectral, they were merged with the slightly coarser spatial resolution (20m) multispectral data (XS) to maximize the information content.

Spatial sampling, however, is not the only parameter influencing the effective spatial resolution of the sensor. Although SPOT-HRV P data are recorded at a nominal 10m sampling interval, the radiance incident on each detector in the CCD array originates from a somewhat larger area. In fact, this area is often two to three times greater than the nominal Instantaneous Field-of-View (IFOV) of the sensor, producing a perceptible 'blurring' effect in the resultant images. The effect is due largely to perturbations induced by scattering of the incident and exitant radiation within the atmosphere. It is often especially pronounced over urban areas, due to the generally higher levels of atmospheric pollution (e.g. higher levels of particulate matter from urban emission sources) over these regions. The effect is not confined to data from the SPOT-HRV instruments, but affects all satellite sensors. For instance, the KVR devices on board the Russian KOSMOS series of satellites have a nominal spatial resolution of 2m–5m: comparisons made between KVR and SPOT-HRV Pan data, however, indicate that the improvement in the visual quality of the former over the latter is not a linear one (in crude terms, the images are not 'twice as good'). Even so, as KOSMOS images become more readily available, they will undoubtedly form an important data source for urban studies.

There are two other aspects of sensor spatial resolution that require consideration, namely

1. the impact of spectral mixing within a pixel, and

2. the ability to derive quantitative information on the morphological properties (i.e. size and shape) of discrete land parcels.

Many image-analysis techniques attempt to distinguish between and to identify discrete objects within the observed scene on the basis of their spectral reflectance properties. Where a single pixel covers two or more spectrally distinct land-cover types, however, a mixed spectral signature will result. This may or may not appear similar to the spectral response of the dominant land-cover type within that pixel, depending on the relative proportions of the different land-cover types present and their individual spectral reflectance properties. In general, a pixel will only have a 'pure' spectral signature (i.e. representative of a single land-cover type) where the land parcel centred on that pixel is at least twice the size of the sensor's IFOV. This implies that, for merged SPOT-HRV P+XS data, objects smaller than ∼30m×30m are unlikely to exhibit a pure spectral response. Consequently, special care is required when analyzing images acquired over urban areas, since these typically have a very complex spatial structure. It assumes even greater significance where the materials used in building construction are derived from local sources and are therefore likely to be spectrally similar to bare soil and rock outcrops in the surrounding area.

The second issue outlined above concerns the analysis of the morphological properties of discrete land parcels within the scene, perhaps to assist in the identification of the dominant

Table 12.1: Example of a land-cover classification scheme (Marrakech).

1. Water, shadow	9. Asphalt, ballast
2. Vegetation	10. Trees
3. Irrigated vegetation	11. Waste lands
4. Crops (various types)	12. High-density housing
5. Exposed soil (dark)	13. Medium-density housing
6. Exposed soil (light)	14. Mixed housing and gardens
7. Highly reflective soils & pebbles	15. Gardens
8. Concrete	16. Undefined

land use in that region or simply for mensuration purposes. In this context, it should be noted that severe problems can arise where the objects of interest are similar in size to the spatial resolution of the sensor, and that mensuration is hazardous unless the objects at least twice this size. Given this, individual buildings can usually not be consistently and accurately delineated in images acquired by the SPOT-HRV sensors in either XS or P modes. Depending on their size and spectral contrast with the surrounding area, some buildings may be identified, while others may not. On the other hand, while individual houses cannot be always be identified, groups of houses and city blocks can often be delineated and, in many instances, interpreted in satellite images.

12.3   Land Use Analysis

12.3.1   Spectral Classification

Digital images acquired by satellite sensors are frequently processed using supervised, multispectral classification algorithms in order to derive thematic maps. In some cases these can be a reasonable substitute for conventional land use maps (i.e. ones based on detailed ground survey). Unfortunately, the relationship between many types of land use associated with urban areas and their spectral properties in satellite sensor images is, at best, complex and indirect (see also Barnsley *et al.*, Chapter 7). In such circumstances, it may only be possible to generate a land-cover map using per-pixel, image classification techniques. To illustrate this point, consider the spectral confusion that can occur as a result of either the same physical material being used in the construction of two very different urban features, or two surface materials, with very different physical characteristics, having similar spectral properties (taking into account the spectral and radiometric characteristics of the satellite sensor). In the first instance, one might envisage considerable ambiguity between an asphalt-surfaced car park and a multi-storey building for which the roofing material is also asphalt. In the second instance, it is often very difficult to distinguish between concrete surfaces in shadow and water, or between bare soils and buildings constructed from compacted-earth.

We have tested several conventional methods of supervised image classification and have proposed a number of alternative approaches (Baudot 1990, Baudot 1995). While in one case — a study of the city of Marrakech in Morocco — it was possible to use supervised image classification to distinguish between as many as 15 different types of land cover (Table 12.1), none was completely successful in solving the intrinsic problem of spectral confusion between some of the fundamental land cover types found in urban areas. The resultant classes are therefore often poorly defined in terms of the real requirements of the end-user. This information may nevertheless be useful for further analysis, but it cannot be considered to be an truly adequate end-product.

Further information on urban areas, for instance on land use, may be obtained through an analysis of the spatial arrangement of the principal objects/land parcels present within the observed scene. At present, however, automatic structural (or syntactic) pattern-recognition techniques are at a comparatively early stage of development (see Barnsley *et al.*, Chapter 7 this volume). While they are often able to recognize elementary texture and patterns, they still fall short when dealing with the much more complex textures and structures that are characteristic of real urban areas. The level of research currently being devoted to this topic gives us considerable hope for the development of operational algorithms in the near future. For the present, though, visual interpretation is still perhaps the most efficient way to extract the necessary information, particularly as far as urban analysis is concerned. In this context, visual interpretation of digital, multispectral images is based on an evaluation of tone and colour, as might be expected, but special emphasis is placed on the use of textural and structural information.

In a study of Marrakech (Baudot 1992), we focused visual interpretation on classes for which the spectral classification yielded relatively low accuracy values (Table 12.2). Thus, large institutional (service) buildings were identified directly from the image data, while their actual function was determined from external sources (e.g. city maps, telephone directories and field survey; Figure 12.1). Large 'unbound' areas with single, specific functions (e.g. cemeteries, military zones and golf courses) were also delineated in this way. Obviously, the cues employed have to be adapted to the specific land use categories found in the city being examined. The most important part of the interpretation process is segmentation of the observed scene into discrete, homogeneous spatial units. The delineation of the boundaries of these homogeneous areas, as well as a coarse determination of their dominant land use, was performed using a merged SPOT-HRV P and XS false-colour composite image, while information on their content was obtained from large-scale, oblique aerial photographs and field survey. Delimitation of the alluvial plains was added to remove any spectral ambiguity that might exist between open water bodies and shadows cast within the urban area. Each area was subsequently associated with a concise description of urban typology (e.g. building type and size, number of storeys, and estimated age) and with a population-density value estimated from aerial photographs and field survey.

12.3.2   Combined Analysis

It has been noted that many urban planners and managers found that the information provided by the land cover maps derived from remotely-sensed data was of relatively low significance to their day-to-day activities, although they appreciated the fine spatial detail that it offered (Baudot 1990, Baudot 1995). On the other hand, visual segmentation of the image into homogeneous land-use parcels was felt to meet many of their information needs,

Table 12.2: Urban categories observed by visual interpretation (Marrakech).

1–3.	Modern housing (3 different densities)	17.	Schools
4–7.	Traditional housing (4 different densities)	18.	Hospitals
8–12.	Unplanned settlements (4 different densities)	19.	Other services
13.	Main road network	20.	Hotel
14.	Main rail network	21.	Gardens of hotel
15.	Alluvial plain	22.	Cemetery
16.	Industry (and related out-buildings)	23.	Military zone

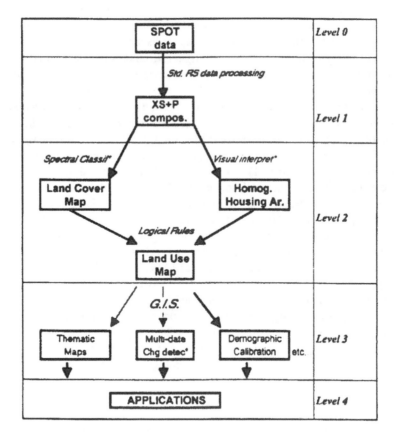

Figure 12.1: General flow chart of the data-processing chain.

although the accuracy with which the parcels were delimited was insufficient. By merging both documents (i.e. the land-cover map and homogeneous housing areas), however, it was possible to combine their respective strengths. The resulting classes were judged to be more compatible with the perceived user requirements than either data source alone. The task of combining the two data sets was achieved using a set of logical rules to recode the large number of potential combinations of land-cover classes and housing area units in a few meaningful ones. For example, a pixel of 'woody vegetation' (land cover classification) in a 'residential' area (photo-interpretation) was re-classified into the 'tree-garden' land-use class. Likewise, a pixel identified as a 'highly reflective mineral' in the land-cover classification was re-classified into the 'urban building works' land-use category if located was in a 'residential' area, or into the 'saline soils and pebbles' category if was located in an undeveloped area.

12.4 Information about the Population

12.4.1 Traditional Sources

Cities are sometimes described as a concentrations of people. Indeed, most of the facts (and problems) about urban areas are related to their inhabitants. Precise knowledge of the population is, therefore, mandatory for almost any management or planning operation. This requires information on both basic (e.g. how many inhabitants are there and where do they live?) and more specific (e.g. water consumption, income, and education requirements) issues. Exhaustive, demographic censuses traditionally form the main source of this information. Unfortunately, the organization of a full national census is so complex and expensive that, in many countries, even the decennial survey recommended by United Nations is not always achieved (Brugioni 1983). In addition, the accuracy of official censuses is known to be highly questionable. Moreover, when the census results do become available (usually 2–3 years later), they may already be obsolete, especially in countries for which annual growth rates of 10% are not unusual. Consequently, there is a need for an alternative, fast, inexpensive and reliable source of demographic information. This is usually provided by sample inquiries. This approach, however, can only provide an estimate of a variable — for instance, the average number of children by household (Cochran 1977). To extrapolate these estimates to the whole population (i.e. how many children are there in the city or country as a whole?), information on the total number of households is also required. Unfortunately, this is often difficult to obtain, except perhaps by means of remote sensing.

12.4.2 Using Remote Sensing for Population Estimation

Conventional aerial photographs have been used for many years to provide valuable information in some countries about human settlements (Adeniyi 1983). They are commonly used to define the boundaries of census tracts, but they can also be employed to make direct estimates of population densities (by counting the number of dwelling units per unit area of land) and even to infer socio-economic information from an analysis of residential typologies. The latter approach is based on the hypothesis that the populations living in areas showing near similar housing conditions will have homogeneous social and demographic characteristics. Consequently, an estimate of the total population within a homogeneous housing district can be obtained by estimating its average population density, and multiplying this by its areal coverage. Naturally, this approach requires recent aerial photographic coverage in fast-growing cities. Unfortunately, in most less-developed countries, where the rates of population growth are often greatest, recent coverage is more often the exception than the rule. Moreover, the organization of dedicated over-flights (where these are possible) will tend to inflate the budget of a method that is intended to be inexpensive.

Satellite remote sensing presents obvious advantages: potentially, data are available for any city in the world, even for foreign analysts. Existing archives can be used to study change over time; their cost is reasonable, if not negligible. It should be evident, however, that given all of the limitations mentioned the preceding section (12.4.1), satellite sensor data cannot provide all of the information required about the characteristics of the urban population. An operational approach must, therefore, involve complementary information sources. In this context, we have developed and applied two approaches to our test sites. The first relies heavily on the satellite sensor data, with only a few field checks to provide a rough estimate of the number and spatial distribution of the inhabitants in a city, and a few indicators that are used to estimate a number of social parameters. This approach was applied to two Moroccan cities (Marrakech and Mohammdia; the latter in an operational situation). The choice of method was imposed by the difficulty of organizing demographic field surveys in Morocco. Nevertheless, comparisons with independent data sources suggest that the 'uncontrolled' estimates are highly credible. The main limitation of this method lies in its inability to provide information on the confidence interval of the derived estimates.

The basic method outlined above was taken a stage further in a study of Ouagadougou (Burkina Faso; approximately one million inhabitants in 1993). In this instance, the satellite sensor data were used as the spatial base around which to organize the complete demographic survey. In this approach, the data processing tasks applied to the remotely sensed images represent only a small fraction of the full set of operations employed. This application was initially envisaged as a methodological pilot study, but its main findings were then employed in various operational applications that were defined by local users.

12.4.3   Method of Sampling

The survey is based on a two-stage stratified sampling scheme (Figure 12.2). The city is divided into discrete strata based on the urban typology defined using the merged SPOT-HRV P and XS images. In each stratum, a sample of households to be interviewed is drawn. A two-stage sampling scheme is used in order to reduce travelling costs (Dureau *et al.* 1991). In it, we select a few sites (first stage) and, within each site, a selection of land parcels (second stage) to be examined. For each site, we take a vertical aerial photograph using a small-format camera mounted on a light aircraft. The location of the photograph is controlled and recorded using dedicated navigation software running on a laptop PC linked to a GPS receiver. These photographs serve as means of locational support during the more intensive field survey, but are also useful in determining the land parcel sizes used in the population density computations.

A questionnaire survey is then administered to selected households. This covers various topics including demography, equipment, revenues, consumption and mobility. Variances

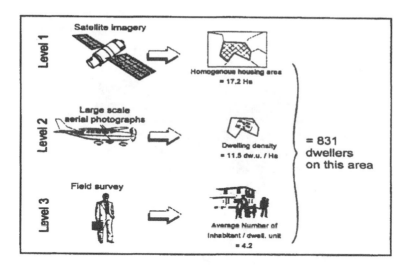

Figure 12.2: Overview of the population estimation method

observed at both sampling levels allow confidence intervals around the estimated mean values to be computed. The level of accuracy obtained differs according to the variables that are studies, but was approximately 90% (at the 0.95 confidence level) in this pilot study. Simulations show that this sampling scheme, optimized according to the observed stratum's variances, should reach an accuracy of 95% with a sample of about 800 households grouped over 200 sites. It is possible (and efficient) to optimize the sampling scheme to achieve the highest possible accuracy, or the lowest possible cost, or any compromise desired between these two aims (Cochran 1977). The principle is to adjust the sampling rates at the first and second levels to equalize the expected accuracies: if a stratum is homogeneous, fewer samples are required. More sophisticated sampling schemes, based on cost-modelling, were tested to achieve the optimum cost-efficiency ratio. However, this approach complicates the practical organization of the survey, sometimes for only marginal improvements in overall accuracy. In practice, we suggest that one part of the available budget is used for a preliminary survey without optimization (proportional allocation). The sampling scheme employed in the second part of the survey can then be adapted to take into account the variances observed during the first stage (Neyman's allocation). This approach does not necessarily lead to the best performance-cost ratio, but it avoids the possible errors obtained by a sampling scheme based on incorrect variance estimates. All variables are expressed as densities. Extrapolation to the whole stratum — and thence to the whole city — is straightforward. This is an important advantage of this approach compared with other sampling methods. Last, but not least, as the stratification criterion is spatially well-defined (the strata are well-defined areas), the results of the survey can also be readily mapped. For instance, Figure 12.3 shows a map of the distribution of population in Ouagadougou, using hatching densities proportional to population densities.

Figure 12.3: Map of population density in homogeneous housing areas for Ouagadougou, Burkina Faso, 1993.

12.5   Using Demographic Information for Land-Use Management

12.5.1   The Concept of Spatial Disaggregation

The two methods described earlier (the rough estimate provided by calibration of homogeneous housing areas and the complete demographic survey based on a stratified satellite-sensor image) provide an estimate of the number of people living in homogeneous areas and, in the second case, a range of more complex socio-economic indicators (e.g. income, consumption, and age structure; Figure 12.3). These results pertain to discrete, homogeneous areas within the city under investigation. When displayed as maps, these results are not quite the same as conventional cartographic products based on census tracts, but they are usually closer to reality than those based on administrative segmentation which is often somewhat artificial. Per-area mapping is obviously one of the most widely-used forms of thematic cartography, but it isn't always the most efficient. Comparisons with other maps based on different area segmentations enjoins the application of area weightings that are not always realistic (Openshaw and Flowerdew 1987). Population maps derived from remotely-sensed data permit a far more efficient method of analysis, namely using the map of homogeneous housing areas and the land-use map to determine the number of pixels allocated to housing land-use categories. In doing so, we implicitly accept the hypothesis that the population within a given homogeneous area is located in the built-up sections of that zone. Accordingly, the total number of inhabitants in a given area, divided by the number of housing pixels in that area, is attributed to each of those housing pixels (where differentia-

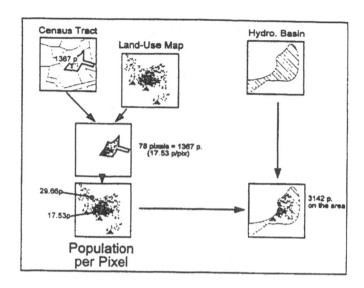

Figure 12.4: Spatial disaggregation of demographic data on housing pixels, followed by spatial aggregation by river basin.

tion between housing types is possible on the classification, adaptive weighting schemes are possible). This process is applied to each housing area, providing a 'population-per-pixel map'.

This kind of document, although not new (see Wright (1936), for example) and sometimes called 'dasymetric map', is not commonly used, but proves very useful indeed (see Donnay and Unwin, Chapter 11, this volume). It gives a precise representation of the actual distribution of the population, but above all, it gives the users the ability to analyze this information after re-aggregating it on the most appropriate spatial base. Figure12.4 illustrates the concept of disaggregation from the information provided on census tract basis to a per-pixel map, followed by a re-aggregation on an area defined by the limits of a hydrological basin. The ability to assess the number of inhabitants on any spatial base is especially important for location-allocation problems. On the test-site of Mohammdia (Morocco), we have computed the number of inhabitants affected by a given concentration of industrial air pollution. On the Ouagadougou test-site, we have demonstrated the potential to determine the amount of population (and related production of dirty water) influencing one hydrological basin, in order to calibrate the construction of a network of sewage and water-purification stations.

12.5.2 Application to Modelling

The potential of population-per-pixel maps was also illustrated for Ouagadougou in a simulation of the optimum location of 10 water-supply terminals for the peripheral districts of the city. In the outskirts of Ouagadougou, fresh water is provided at free-access water

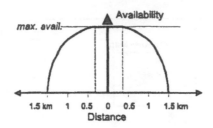

Figure 12.5: Model of decreasing availability used for the location of fresh water terminals in Ouagadougou.

terminals. People living in the close neighbourhood of a terminal serve themselves, while water-carriers supply water to distant locations. The price of the water provided is not related to the distance. Thus, rather than a model of cost, we used a model of availability. People living near an existing water terminal are provided with water directly (maximum availability). As the water-carriers have to supply all possible locations around the terminal with a given amount of water in their tank, the potential availability decreases with the squared distance from the source. One considers that $1.5km$ is the maximum distance allowable for daily usage. The model of availability is thus summarized in Figure 12.5.

The model for the location of water supply must take account of numerous properties. For instance, the model should not allocate multiple, closely-located water terminals in densely populated sites; rather it should indicate the requirement for greater output of water from a single sites in such circumstances. In other words, once a location is selected, the population of its hinterland should not be taken into account again during the selection of other possible water-terminal sites. Information on land use is also important, especially in urban areas, since some locations are simply not appropriate sites for water terminals.

We use the map of population-per-pixel and the model of availability to compute a continuous surface of potential water supply. For each possible site (i.e. for each pixel of the land-use map), the total population within a radius of 1.5 km is calculated, weighted by the distance/availability model. In this particular study of Ouagadougou, only the population located outside the peripheral road are considered. Ten sites of highest potential water supply are selected on the 10 highest local maxima of this surface. The final location of each water-terminal is then refined using the land use map and a digital terrain model, and the potential demand for this location is re-computed. For each site, the total number of inhabitants living within a given distance is determined in order to calibrate the output and infrastructure of each station (Table 12.3). Figure 12.6 shows the result of this allocation.

12.6   Accuracy of the Estimation

12.6.1   Requirement for Quality Assessment

In most instances where spatial data are used within a GIS, an assessment of the quality of these data should be performed. Data quality not only embraces the general accuracy of

Table 12.3: Allocation of the population to the proposed water terminals around the periphery of Ouagadougou.

Site #	$0 \leq d \leq 0.5\,km$	$0.5 \leq d \leq 1\,km$	$1 \leq d \leq 2\,km$	$0 \leq d \leq 2\,km$
1	23399	13294	4223	40916
2	24175	20183	10802	55160
3	18947	11480	72	30499
4	21206	18881	7237	47324
5	17909	5873	186	23968
6	19390	17724	3378	40492
7	17613	22692	4610	44915
8	17232	8226	9650	35108
9	13317	12123	1034	26474
10	9875	1207	3001	14083
Total	183063	131683	44193	358939
(%)	51.0	36.7	12.3	100.0

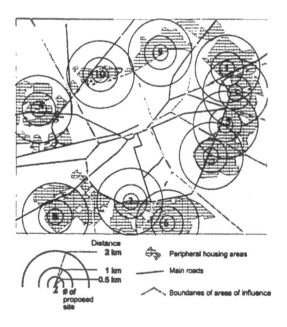

Figure 12.6: Location of 10 water terminals optimizing the supply of fresh water to the outskirts of Ouagadougou.

the input data, but also complete documentation of the lineage of the results output from the analysis of those data. This is especially important when the output data are intended for use in subsequent processing stages. The propagation of errors in GIS models is not trivial or unimportant and should be taken seriously by all GIS analysts. Regrettably, quality assessment is not yet common practice among GIS practitioners — both suppliers and users. In the context of this study, we have attempted, as far as practically possible, to document the quality of our results. Unfortunately, this is not always an easy task.

12.6.2   The Traditional Approach

Classification Accuracy

The population estimate, obtained via the methods described in the previous section, derives from a satellite-based land use classification/segmentation and from a stratified sampling survey. The accuracy of a land use classification is usually expressed in terms of a confusion matrix. This can be used to assess the general quality of the classification, which is sometimes summarized by an overall precision index (percentage of correctly-classified pixels, Kappa coefficient, *etc.*). In terms of the typical classifications schemes employed with respect to images of urban areas, some classes may only represent a very small part of the total area but may nonetheless be very significant in the planning context. In such situations, the normal statistical requirements — i.e. a sufficiently large sample size and the same sampling rate for each class — can place unrealistic constraints on the exercise. Moreover, intensive field survey can often be problematic in some urban areas, where accessibility is inhibited either for cultural reasons (e.g. in traditional housing of Muslim countries) or for security reasons (e.g. in the slum districts of certain cities).

Sampling Accuracy

The accuracy of a random sample is related to the variance of the population and to the sample size. It is usually expressed in terms of the variance of the estimate and is relatively straightforward to calculate even where a relatively complex model of stratification and/or multiple levels has been employed. A problem with this form of accuracy estimation is that it is based on the hypothesis of that the sampled variables are normally distributed. It is not always easy to verify such hypothesis. For our test sites, several of demographic variables do not conform well to the normal distribution model. The effects of such a bias on the resultant quality assessment are difficult to predict.

Other Error Components

In spatial data bases, the position component of the information is also important. If the satellite data are not perfectly geo-referenced, even a properly classified pixel may be incorrectly interpreted due to residual mis-registration. Again, it is not always simple to separate the effects of attribute errors from those related to positional accuracy, especially when very small parcels of land are concerned.

One other spatial aspect relates to the spatial base used in the analysis. Some variables are easily defined in a geographical system. For example, crop types can be related to their associated agricultural land parcels, housing functions to the footprint of a the corresponding buildings, etc. However, it is generally much more difficult to assess the location of the human population: it is by nature mobile (the daytime and night-time distributions may be very different in some districts). Typically, demographic variables are gathered and referenced by household, and households are assigned to a precise dwelling. Unfortunately, these schemes are not always very representative in the context of Third World countries. In many cases, the notion of 'household' is fuzzy, while shared dwellings (multiple occupancy) is common. It is particularly difficult to cope with these kind of inaccuracies in deriving an estimate of global quality.

12.6.3   A More Pragmatic Approach

If the aim is to develop an operational system, the final products have to be conform to the user's requirements. Usually, the user is not interested by the quality of intermediate results such as the land cover/land use classification and the sampling survey, but the possible usage of these data may be of value. Most often, the user will prefer a pragmatic expression of the expected accuracy rather than a complex value expressed in statistical terms, even if the latter is more complete. We therefore suggest that users should also be provided with an easy-to-understand assessment of overall quality. For instance, the general accuracy of the final estimated variable (e.g. the amount of fresh water required) might be expressed as the difference observed for selected sample areas (independent of the sample used to derive the estimate) between the values announced by the proposed GIS model and a reference obtained by field inquiry.

12.6.4   The Propagation of Errors

Theoretical models of propagation are available, but their applicability is not straightforward or self-evident where complex GIS models are concerned. The main interest in these models is to highlight the principal sources of errors that contribute to the final results. On this basis, we estimated the relative importance of errors in the initial land use/land cover classification (generating locational errors) and those arising from the demographic sampling (treated as attribute errors), and tested their mutual interaction using Monte-Carlo simulation. The simulation shows that improvements in the accuracy of the demographic estimate (attribute) are more significant in terms of the accuracy of the final product than similar improvements to the land cover/land use classification. To place this evaluation in a strictly operational context, the analysis should also integrate the marginal cost of improvement for each type of error. In this case, the cost needed to improve the accuracy of the demographic field survey is generally much higher than that required to achieve a one percent gain in classification accuracy.

12.7   Conclusions

In developing countries, the growth of cities is startling and demands special attention. Whatever the form of the proposed solution, it will inevitably require efficient diagnostic and simulation tools. It is self-evident that many of the problems facing Third World cities (and hence their solutions) are related to economic, social and political factors, but land-use management cannot be neglected if sustainable development is to be achieved. A city is a concentration of human beings in which the population not only acts on the city, but is also influenced by it. An integrated approach to urban development cannot therefore ignore the human population component.

Remote sensing has proved its ability in various application domains to provide valuable information about land use. Urban environments are often considered too complex to be analyzed by satellite remote sensing and, indeed, the spatial resolution of current satellite sensors means that they are not particularly well suited to the task. On the other hand, in many situations remote sensing is simply the only available source of data. In fast growing cities, if the information is not recent, it is useless. That said, the information provided by current satellite sensors is probably best adapted to the broad-scale analysis of large cities, rather than to detailed investigation of a few city-blocks. Nevertheless, the rising demand for global structure plans for Third World cities augurs well for the application of remote sensing to urban areas in the coming years.

As stated earlier, the human population in urban areas deserves special attention, but traditional information sources often fall short of the basic requirements. While satellite remote sensing cannot provide demographic information directly, when used in conjunction with other data and integrated methods of analysis it has considerable value. High resolution satellite images, in particular, can be used to set up sophisticated demographic sampling surveys. After that, the same data can be used to generate land use information, in order to permit efficient spatial analysis. The potential of such information for urban management seems as wide as the potential of the GIS tools used to exploit it — only limited by the imagination of the users.

12.8 Acknowledgements

The main part of this research was developed during the Belgian Scientific Research Programme on "Remote Sensing by Satellite — Phase Two" (Services of the Prime Minister — Science Policy Office). The scientific responsibility is assumed by the author.

12.9 References

Adeniyi, P. O., 1983, An aerial photographic method for estimating urban population. *Photogrammetric Engineering and Remote Sensing*, 49, 545–560.

Baudot, Y., 1990, A new approach to satellite image classification using a color data compression algorithm, In *Proceedings of IGARSS'90* (Washington, D.C.: IEEE).

Baudot, Y., 1992, Application of remote sensing to urban population estimation: a case study of Marrakech, Morocco, In *Proceedings of the EARSeL Workshop: Remote Sensing and GIS integrated for the management of less-favoured areas*, EARSeL (EARSeL), pp. 138–147.

Baudot, Y., 1995, Applications of remote sensing to the management of urban areas in less-developed countries, In *Space Scientific Research in Belgium, Volume III — Eart*

Observation Part 2 (Brussels: Federal Office for Scientific, Technical and Cultural Affairs), pp. 9–18.

Brugioni, D. A., 1983, The census: it can be done more accurately with space-age technology. *Photogrammetric Engineering and Remote Sensing*, **49**, 1337–1339.

Cochran, W. G., 1977, *Sampling Techniques*, Wiley Series in Probability and Mathematical Statistics, third edition (New York: John Wiley and Sons).

Devas, N., and Rakodi, C., 1993, *Managing Fast Growing Cities* (Harlow, Essex: Longman).

Dureau, F., Barbary, O., Michel, A., and Lortic, B., 1991, *Sondages aréolaires sur images satellite pour des enquêtes socio-démographiques en milieu urbain (15 fascicules)* (Paris: ORSTOM).

Openshaw, S., and Flowerdew, R., 1987, A review of the problems of transferring data from one set of areal units to another incompatible set, Research Report 4, Northern Regional Research Laboratory, U.K.

Wright, J. K., 1936, A method of mapping densities of population with Cape Cod as an example. *Geographical Review*, **26**, 103–110.

PART V

EPILOGUE

CHAPTER 13

Remote Sensing and Urban Analysis: A Research Agenda

Paul A. Longley, Michael J. Barnsley and Jean-Paul Donnay

13.1 Challenge to Remote Sensing Analysis

The European Science Foundation meeting on which the chapters of this book are based took place at an interesting and exciting period in the development of urban remote sensing. In some senses, the meeting came too early in that it predated both the launch of the 'new generation' of very high spatial resolution (optical) satellite sensors, and the wide dissemination of a variety of digital data sources that can now be used to augment detailed satellite data. Some of the techniques presented at the meeting were proprietary, and their natures were neither immediately transparent nor yet open to scrutiny through the peer-reviewed academic literature. Opaqueness of exposition survives in some of the contributions to this book, although we have tried to make the message, if not the detail, of the contributions as clear as possible. In these respects — paucity of 'good' data and a 'grey' literature of developing technique — the formative days of urban remote sensing (RS) seem curiously akin to the early development of GIS. Today the field is developing rapidly (Donnay *et al.*, Chapter 1 this volume) and is converging with mainstream GIS applications (Atkinson and Tate 1999). As such, there are now very real prospects that 'RS-GIS' can provide a near seamless software environment for urban analysis.

The uses of data and technique in some of the contributions to this book have a formative rather than a strictly contemporary feel, yet the collection provides a still prescient evocation of thinking about the measurement and analysis of urban systems. The contributions convey a sense of achievement that remote sensing is on the threshold of creating really useful urban inventories, tempered with some disquiet that increased technical precision in detecting the extent and morphology of urban 'land cover' begs at least as may questions as it resolves.

Specifically, there is a realization that better spectral detection of land cover using improved optical instruments provides only the most indirect of indicators of urban 'land use'. Whilst the promise that further improvements in the precision of satellite instrument measurements might yield 'better' data, there is a realization that there is no immutable chain linkage to 'better' urban analysis. Further progress requires reappraisal and rethinking of various aspects of urban theory, the shaky foundations to which are already now partially exposed. In particular, it is increasingly apparent that there has been little clear thinking about the size, shape, scale and dimension of elemental units that might be used to define the 'density' of urban land uses. Clear conception of phenomena logically precedes their measurement in science, yet the first forays into urban remote sensing have apparently been driven by the reverse logic. Measurement usually precedes theory in science too (Mandelbrot 1982), and thus remote sensing is pivotal in terms of both forward and backward linkage to our conception and understanding of urban systems.

The nature and pace of technical developments in remote sensing was, and remains, very impressive (Donnay *et al.*, Chapter 1 this volume), yet conceptual issues are of very much more than semantic importance if the science of remote sensing is to make significant contributions to the rational planning process. Remote sensing is a technology that has hitherto portrayed rather conventional and objective measures of spatial distributions, yet the detailed interpretation of urban form clearly needs to move beyond conventional, per-pixel classifications of detected spectral reflectance. Configuration, syntax, structure and function need to supplement conventional spectral, spatial, temporal, geometrical and polarization clues to land cover (Curran *et al.* 1998, Barnsley 1999). Indeed, our understanding of morphology needs to go beyond the physical to the socio-economic, since similar built structures and site layouts can fulfill a range of diverse socio-economic functions. A geography of the inert carcass of the city, however intricate, is of only limited usefulness to planning for human activities. Remote sensing representations of urban systems are geographically comprehensive, frequently updated and increasingly detailed at large scales. Yet, at the end of the day, they provide only a physicalist conception of space, which needs to be augmented by other sources if they are to make significant contributions to our understanding of urban systems. If this can be accomplished, there is no reason why urban remote sensing should not make an important contribution, not just to understanding data, but to understanding urban systems, and not just in terms of their form, but also their functioning.

The remote sensing tradition has focused very much on technical issues of data assembly and physical classification. Yet the range of contributions to this volume suggests an emergent inter-disciplinary focus, with interest from planners, architects and geographers concerned with specifying context in order to ascertain detail. In this concluding chapter we will try to develop this view of urban remote sensing as an inter-disciplinary meeting place, where we can assess the way that data inform our understanding of the spatial distributions of urban phenomena, and how these, in turn, may inform practical planning issues.

There is certainly a need for this, and the spatial and temporal properties of urban remote sensing data are pivotal to a systematic attack on a wide range of planning problems. The classifications of urban form that are created by urban remote sensing are spatially comprehensive, and this presents renewed opportunities for analysis of the 'city as a system within a system of cities' (Berry 1973). Within developed countries, the locus of planning debate in recent years has shifted towards encouragement of 'sustainable' cities, predicated upon the reuse of 'brown' land. The kinds of measures proposed by Mesev and Longley (Chapter 9 this volume), for example, provide a means of using satellite imagery to identify the way that urban development fills space, and by implication suggest ways in which urban morphology might be manipulated to further urban policy objectives. The spirit of this work is in the established locational analysis tradition of human geography (Haggett *et al.* 1977) which has already been developed, for example, to investigate whether development restrictions (such as UK 'green belt' policy) change the shape and scale of urban areas (Longley *et al.* 1992). Informed by better data, the class of cellular automata models may be used to extend 'what is' models of detailed distributions to 'what if' scenarios of urban growth and change (Wu 2000).

Within the temporal domain, many of the pressures upon urban planning are generated by the sheer scale and rapidity of economic and social restructuring of cities and their consequent effects upon spatial structure. The rates of growth in the cities of developing countries are unprecedented, and the urban data infrastructure for monitoring them is not nearly as well developed as that in the West. New technology is creating a sea change in the way we think about infrastructure, as in the use of digital mobile phones in terrain where no land lines exist, or the use of global positioning system (GPS) receivers to map large tracts of the developing world. It is in this spirit that Baudot (Chapter 12 this volume) describes how urban remote sensing provides a means of monitoring changes in the extent and form of urban settlements in the developing world. In parts of the world such as North and West Africa, conventional urban data sources at best provide a normative, official 'what should be' inventory of ownership interests rather than a 'what is' representation of formal and informal urban land use.

The challenges in the developed world are rather different. Here, national populations are more or less constant in size, although their residential preferences and propensities to form and dissolve households are manifest in a range of different tensions in the urban fabric. The diverse lifestyles and living arrangements of urban populations are forcing restructuring of the built form of cities, such as the 'densification' of established residential areas, the decentralization of retailing, changes in industrial location patterns and so forth. Here, urban remote sensing complements what is available elsewhere, and presents a valuable, up-to-date and comprehensive picture of the changing extent and morphology of urban areas.

Taken together, these contextual issues suggest a challenging role for the science of remote sensing in social science formulations of the form and functioning of urban systems. From a technical standpoint, it is clear that urban remote sensing is developing for two primary reasons:

1. new sources of satellite data; and

2. vastly increased computer power, new methods of geocomputation, and new thinking about science through induction.

To these must be added two further sets of considerations, which are prerequisites to sustained development of the field:

1. the development of new digital data infrastructure, and its use as ancillary sources for RS image classification; and

2. reappraisal of the ways that an understanding of data fosters improved understanding of urban systems.

13.2   Understanding Representations of Urban Areas

The science of remote sensing is fundamentally concerned with obtaining information about the physical, chemical and biological properties of the Earth's surface (Barnsley 1999). While some of these properties can be estimated using statistical (empirical) or physically-based models, most are currently inferred by less direct means, often by grouping image pixels into discrete categories using multispectral classification algorithms. Nowhere is this more apparent than in the study of urban areas, where the production of land use maps and the estimation of population counts is generally predicated — wholly or partially — on some form of image classification. Such categorization is suggestive of a much more 'black-and-white' view of the world than typically characterizes either natural or man-made systems. Fisher (1999), for example, describes the inherent vagueness and ambiguity associated with such apparently discrete characteristics as 'oak woodland', in which the frequency of occurrence of the defining characteristic or the degree of membership of a specific environmental category may vary across time and space. More generally, geographers have long reflected the paucity of 'natural' units in the real-world (Openshaw 1984): in this context, it will surely be only coincidence if the most appropriate unit to measure land-surface properties in a particular remote sensing application happens to coincide with the dimensions of the image pixel. Taking this logic a stage further to consider appropriate size and scale thresholds for urban land use classification, we encounter additional problems of definition, since 'urban' is very much a 'catch all' phrase, and urban forms conceal a multitude of functions.

Elsewhere in urban analysis it has become common practice to adopt working definitions of 'urban' which are suited to particular purposes. In the early days of GIS, for example the (then) UK Office of Population Censuses and Surveys (OPCS 1984) created vector digitized boundaries of irreversibly 'urban' land in accordance with contiguity and population size criteria. The principal purpose of the national dataset was, quite simply, urban inventory analysis, although Longley *et al.* (1998) subsequently demonstrated that even measures such as these could be used to establish allometric relations between size and area across a regional settlement system. Such scale relations also enable anomalies to be highlighted, particularly in a scattered system of small settlements which is not greatly differentiated by function. Yet for a wider range of applications, the boundaries are crude, the definitions upon which they are based are inadequate or inappropriate, and they imply uniform population and land use densities within the urban boundary. Ten years later, the interdependence between measurement and outcome is resurfacing in urban remote sensing, as the contribution of Weber (Chapter 8 this volume) illustrates.

These dilemmas extend well beyond the realm of urban inventory analysis. In the early days of quantitative geography, analogies in terms of shape and form were drawn between cities and biological organisms (Haggett *et al.* 1977, Batty and Longley 1994). Similar linkage between city genealogy and biological Darwinism is developed by Pesaresi and

Bianchin (Chapter 4 this volume), with the additional connotation that the whole of a settlement is greater than the sum of its parts. Of course, this is but one framework for exploring the functioning of individual urban systems, yet it provides an illustration that understanding city morphology requires unambiguous and precise measurement of the functional elements of the system. It also implies a 'horses for courses' approach to data creation, which sets urban data modelling in stark contrast to the traditions of remote sensing.

An agenda for urban remote sensing should therefore accommodate the need to make data classification sensitive to purpose. This is also the case in inter-urban analysis, where urban geographers have long been involved in the search for order in urban systems, with Christaller's Central Place Theory providing perhaps the best known example. Analysis of settlement hierarchies also implies an understanding of the functioning of regions, since the threshold for a particular urban function does not necessarily correspond to spatial extent of land coverage, as any visitor to Hong Kong or Singapore will testify. The spatial manifestations of urban function are partial and imperfect. It follows that we need ancillary sources to ascertain function, and that our choice will be related to the kinds of function that we are investigating.

The responses to date within urban RS classification have been essentially pragmatic. They have *imposed* size, shape geometry and scale constraints upon classifications, all in an essentially *ad hoc* manner. Bähr (Chapter 6 this volume), for example, casts his classification of urban land cover into a Delaunay Triangulation to determine the extent of an urban area. Similarly, both Bähr (Chapter 6 this volume) and Weber (Chapter 8 this volume) adopt spread and containment functions to augment their urban land cover classifications of optical and microwave data, although their specific solutions differ in detail and purpose. Bähr, for instance, contends that multispectral images over-estimate urban areas, while Weber deems an additional 200m buffer zone necessary to delineate urban areas using classified SPOT data. The stark contrast between these different empirical, data-specific 'fixes' to the task of urban image classification highlights the subjectivity still embedded in this process. Bähr's second case study also sets an arbitrary minimum size threshold for an 'urban' area, while both Pesaresi and Bähr take steps to classify isolated parcels of urban land use even though these are unlikely to fulfill any urban function beyond shelter.

While it seems intuitively reasonable to expect urban land classifications to be 'fuzzy', there is clearly some need to investigate the plausibility of different spread or containment functions, and of any imposed land use geometry. This also raises a further important issue that has been known for some time in computer cartography: namely, how much operator intervention in the classification procedure is necessary and/or desirable? There is a point beyond which further operator intervention diminishes the generality of urban land use classifications. This discussion suggests a number of supplementary issues that must be addressed in the classification of urban land cover/use. First, the arbitrary spread and containment functions that have been used in classification work to date represent a first fumbling towards incorporating a syntax of space that is more sophisticated than elemental 'geographical bits' (e.g. pixels). Second, spectral data pertaining to land cover need to be supplemented with ancillary information (e.g. about population size) to yield plausible classifications of land use. These, in turn, raise broader issues about geocomputation, ancillary sources of data infrastructure and the scientific practice of image classification.

13.3   Data Availability and Advances in Data Processing

The development of remote sensing has very much been technology-led — notably through improvements in sensor design and advances in the supporting computational infrastructure. Issues relating to developments in sensor technology were addressed in the introduction to this volume (Donnay *et al.*, Chapter 1 this volume). At the same time, precipitous falls in the real and absolute cost of computing have made it possible to store, manipulate and analyze greater volumes of data, more rapidly and, arguably, with greater ease than ever before. This has been central to the continuing development of satellite remote sensing, including that pertaining to urban areas, which now generates some of the largest data sets in the environmental sciences and almost certainly dwarfs all social survey sources.

In the context of urban analysis, however, the increasing availability of new sources of digital, socio-economic data sets may be considered equally, if not more, important. These derive from the everyday interactions of individuals with computers, not just through digital data capture of social surveys, but also from address registers and 'traces of activity' patterns, such as ATM (Automatic Teller Machine) and EPOS (Electronic Point Of Sale) transactions. Technology has played no small part in the fission of urban lifestyles in the post-modern city, but it also empowers us to track and monitor the scale and scope of these changes (Longley 1998). As a result, digital depictions of geographical reality are increasingly capable of capturing the physical and socio-economic structure of cities, even though the best (computational and statistical) means of achieving this is still subject to debate. More broadly, the challenge to what has become known as 'geocomputation' (Longley *et al.* 1998) is to concatenate and conflate a wide range of diverse geographical data sources with different real-world structures.

While advances in satellite sensor technology, particularly those relating to sensor spatial resolution, are helping to make remote sensing more appropriate to the study of urban areas, parallel developments in image classification techniques — notably, artificial neural networks (ANNs) and fuzzy set methods — may offer the prospect for improved representations of the urban fabric (Atkinson and Martin 2000). Clearly, these techniques have not been developed specifically with urban remote sensing in mind, but they are highly appropriate to the problems associated with images of such areas (i.e. uncertainty associated with object boundaries, fuzziness in terms of class definitions, and the potential for any given land parcel to be a member of more than one land cover/land use category). Neither of these techniques is represented in the chapters of this book, partly due to the timing of the ESF meeting, but they are becoming increasingly standard practice. Each offers benefits with respect to standard multispectral classifiers, such as the maximum likelihood algorithm, which has been used extensively throughout this volume. In the case of ANNs, these benefits include:

1. the need to make fewer assumptions about the statistical distribution of the input data;

2. the ability to generalize on the basis of the patterns presented to the network; and

3. a degree of tolerance to noise in the training data (Foody and Arora 1997, Mather 1999).

On the other hand, deriving accurate classifications from ANNs is arguably still more of an art than a science, with critical decisions needing to be made with respect to the most appropriate network architecture and its configuration in terms of the number of nodes and hidden layers, as well as the optimum sample size and the number of iterations needed to train the network (Blamire 1996, Paola and Schowengerdt 1997). Fuzzy classifiers are, in some senses, even better adapted to the needs of urban remote sensing, since they deal explicitly with the problem of uncertainty in object (e.g. building) boundaries, as well as multiple and partial class membership (Fisher and Pathirana 1990, Binaghi *et al.* 1996, Foody 1996, Zang and Foody 1998, He *et al.* 1999). Not surprisingly, there have also been various attempts to merge the two approaches in order to exploit their individual and combined strengths (Lee and Lee-Kwang 1994, Cho and Kim 1995, Foody 1997).

Ultimately, however, while ANN and fuzzy classification techniques may provide more accurate representations of land cover within and around urban areas, they currently offer little more than standard multispectral classification algorithms in terms of solving the more taxing problem of determining urban land use from remotely-sensed images. Despite their actual or perceived computational advantages, the practical application of these new techniques remains stubbornly focused on per-pixel characterizations of land cover (although see Atkinson and Tate (1999) for some illustrations of polygon and object classifications). If we are to move beyond this, to representations of urban land use, new ways must be found to measure, model and eventually understand the syntax of urban space and the configuration of urban social areas.

Two approaches to this broad problem have been posited in this volume: one empirical, based on statistical measures of image texture/structure; the other model-based, employing techniques drawn from the field of syntactic (or structural) pattern recognition. The former is exemplified by the work of Brivio *et al.* (Chapter 3 this volume), who employ geostatistical techniques — specifically the semi-variogram — to characterize the spatial structure of urban areas. Their work confirms expectations that different types of urban land use exhibit measurably different spatial patterns in terms of their reflectance properties in high spatial resolution remotely-sensed images. They also demonstrate that the relationship extends to towns and cities built in different eras. Although Brivio *et al.* do not attempt to invert this relationship to derive maps of land use from statistical measures of the image structure, the potential application of what might be referred to as 'geostatistical image classification' is clearly raised. Pesaresi and Bianchin (Chapter 4 this volume) employ a related empirical approach, based on measures of image texture and mathematical morphology operators, to identify the extent of urban areas across a large section of northern Italy.

An alternative, model-based approach is adopted by Barnsley *et al.* (Chapter 7 this volume), who adapt aspects of syntactic pattern recognition to meet the needs of urban remote sensing. Although structural pattern recognition has been used extensively in the field of machine vision — including automated analysis of aerial photography for military purposes — it has not yet been widely applied to the analysis of civilian satellite sensor images of urban areas (Barr and Barnsley 1999). This situation seems set to change with the advent of the new generation of very high spatial resolution (<5m) satellite sensors. These are better suited to the detection of individual objects (e.g. buildings, roads and various types of open space) within urban areas, and hence to explore the spatial and structural relationships between them. Like Brivio *et al.*, Barnsley *et al.* demonstrate credible relationships between various categories of urban land use and the structural patterns evident in remotely-sensed images, but have yet to develop data-processing techniques that can be used to invert this relationship (i.e. to map urban land use as a function of the spatial patterns observed in the image data). They note that the problems involved in resolving this are non-trivial.

Despite the technical advances embodied in the work reported by Brivio *et al.*, Pesaresi and Bianchin, and Barnsley *et al.*, a number of fundamental questions remain unanswered. Many of these arise because the research agenda has been predominantly data-driven — answering pragmatic questions akin to "what can be obtained from existing image data sets"? There has, for example, been little or no theoretical consideration of the spatial resolution (scale) of image data most appropriate to statistical or structural pattern recognition. Similarly, there is as yet no formal theoretical basis to establish the principal classes of object (e.g. buildings, roads and various types of open space), and the structural relationship between them, that are needed to characterize different categories of urban land use. There are also questions about the consistency of these properties and relations within and between different urban areas. Undoubtedly, hypotheses will start to be formed from the types of empirical investigation reported in this volume, but inductivism may only take us so far. Moreover, there are limits to what can be achieved through an analysis of remotely-sensed data on its own. Consequently, in the next section, we will turn our attention to the broader digital infrastructure and the application of ancillary data sets.

13.4   Digital Infrastructure and Ancillary Data

One of the great strengths of urban remote sensing is that straightforward application of standard classifiers to optical imagery can yield seamless digital classifications of artificial built structures which are both geographically extensive and comprehensive. Such classifications may be enhanced with reference to data from other optical survey instruments (Ranchin *et al.*, Chapter 2 this volume), from microwave sensors (Bähr, Chapter 6 this volume), or from repeated measurements through time (Bähr, Chapter 6 this volume). Yet most of the legitimate questions (Donnay and Unwin, Chapter 11 this volume) which are asked in urban analysis require us to think of the city not just as a built structure, but also as the locus of human activity patterns. The challenge, therefore, is to use remote sensing classifications as one of several inputs to an analysis of the ways in which humans interact with the built environment.

Within developed countries there are other sources of information which are conventionally used to answer this type of question, including (i) the Censuses of Population carried out, among other places, in the US and UK, (ii) address and property registers, and (iii) standardized NUTS (Nomenclature of Statistical Territorial Units) data across the European Union. Yet such data variously exhibit a number of shortcomings. These include:

- outdatedness — the decennial snapshot interval of the US and UK Censuses, for example, is too infrequent to monitor fast-changing urban development trends,

- coarse zonation — necessary to preserve confidentiality, yet effectively degrading the resolution of data to that of crude choropleth maps, and

- incompleteness/irrelevance — information is only on available domicile, not work-place and leisure activities.

In short, most conventional government data series bear little correspondence to the de-tailed morphological dynamics of urban growth and change. Some of these problems can be resolved through use of new sources of 'framework' data, created by national mapping agencies to provide precise coordinate references for built structures. These include ATKIS in Germany (Bähr, Chapter 6 this volume) and ADDRESS-POINT™ in the UK (Longley and Mesev 1999). Some framework sources enable land use to be deduced from the address labels that are ascribed to coordinate pairs. For example, ADDRESS-POINT™ records the centroid of every individual building that is a mail delivery point to sub-metre precision. The properties of this type of framework data vary according to the institutional structures in different national settings (Rhind 1997, Guptill 1999). Some commentators (Bähr, Chap-ter 6 this volume) see them as having a symbiotic relationship with remotely-sensed data of urban areas. In this view, the frequency with which remote sensing data are obtained makes them a potential source of updating information (although post office data can fulfill a sim-ilar function), while framework data present a means of ascribing micro-scale structure to the built forms detected by high resolution imagery.

The range of socio-economic and framework sources that is now available provides a powerful means of enhancing urban remote sensing classifications (Mesev *et al.*, Chapter 5 this volume; Longley and Mesev, Chapter 9 this volume). The use of small area census data to augment standard image classification procedures is now a fairly standard procedure in urban remote sensing (Donnay and Unwin, Chapter 11 this volume; Mesev *et al.*, Chapter 5 this volume), and can be used to devise improved measures of space-filling (Longley and Mesev, Chapter 9 this volume). In one sense, the precision of some of these new framework sources (notably ADDRESS-POINT™) make them an effective substitute for remotely-sensed data of urban areas, particularly with respect to the densities of built structures in different land use classes. Yet, as Longley and Mesev (1999) have shown, the spatial object transformations inherent in reducing three-dimensional built structures to point representa-tions create a raft of new problems regarding the way that development appears to fill space. The most obvious solution to this is to ascribe arbitrary dimensions to each georeferenced point, although this obscures geographical variability in the nature of the built form.

The kinds of abstractions provided in framework data force us to reconceptualize what we mean by terms such as 'density'. In a partial sense, the new generation of digital frame-work data products also resolves ecological fallacy and modifiable areal unit problems in spatial analysis (Openshaw 1984). The most rapid progress in urban analysis will be made if the best models are based upon the best data, although many are marketed as commer-cial products and are no more in the public domain than some of the proprietary solutions described elsewhere in this book.

13.5 Analysis

In general terms, spatial analysis has been defined as 'a whole cluster of techniques and models which apply formal, usually quantitative structures to systems in which the prime variables of interest vary significantly across space' (Batty and Longley 1996). This view of space as littered with objects can be re-expressed to emphasize the importance of the container in which they are placed. Thus spatial analysis is also 'that subset of analytic techniques whose results depend on the frame, or will change if the frame changes, or if objects are repositioned within it' (Goodchild and Longley 1999). It is clear from the discussion of ancillary sources above (and also Mesev *et al.*, Chapter 5 this volume; Longley and Mesev, Chapter 9 this volume) that definition of the 'prime variables of interest' in urban analysis is fraught with ambiguity. It is also clear that spatial constructs such as 'density' are very frame-dependent. As elsewhere in spatial analysis, the conception and measurement of spatial phenomena very much conditions what spatial analysis can be expected to achieve.

This kind of revelation is new to remote sensors who generally posit essentially objective definitions of spatial objects, yet it is not new to urban geographers and planners. During the 'Quantitative Revolution' in geography, analysis of social patterning became focused on a description of the patterning of a mosaic of urban areas, as defined by censuses or other large-scale public sector surveys (Timms 1971). Yet using the environment of GIS to handle the much richer and more disaggregate sources that are available today, Longley and Harris (1999) demonstrate that the high degree of socio-economic heterogeneity within small areas very much compromises the validity of the 'mosaic metaphor' (Johnston 1999) of conventional urban analysis. Thus, aggregation of elemental (human) units into artificial zones has been shown to obscure much of the diversity of socio-economic conditions at the small-area level. Applied to remotely-sensed images of urban areas, large tracts of urban land with similar spectral signatures clearly fulfill diverse social and economic functions. Such building blocks are not likely to enhance the prospects for meaningful spatial analysis.

Thus far we have set out the guiding role for urban remote sensing as allowing us to create detailed inventories of urban land uses — such as residential, commercial, public open space, industrial — by direct or indirect inference from satellite sensor images, either in isolation or in conjunction with various ancillary data sources. This is still very much the realm of inventory creation, although the use of ancillary sources potentially extends the scope to cataloging small-area activity patterns. Such operations represent only a starting point for rational planning analysis, and 'what is' inventory statements need to be developed into 'what if' scenario testing if the full power of data-rich urban representations is to be unleashed.

Urban form has been related to function across a range of scales from the architectural to the intra-urban (Batty and Longley 1994). Yet the nature of the relationship is only direct in exceptional circumstances, where the will of the few imposes rigid geometrical constraints upon the built environment for the many, as in Renaissance fortified towns or the early plans of the Garden Cities movement (Batty and Longley 1994). The experience of history is that urban form is structured, yet irregular, and this is certainly borne out in the way that 'cities of pure geometry' such as Karlsruhe break up with distance from the planned central areas, as preordained street layouts become overwhelmed by the many and multifaceted forces operating in the urban land market. Urban theory is predicated on the notion that residential neighbourhoods, retail systems and commercial functions operate according to strict hierarchical principles. These are self-evidently not manifest in pure geometrical arrangements, and the kinds of idealized city structures that adorn high school textbooks are not apparent in the world around us. The kinds of thematic maps produced from classifications of remotely-sensed data are much more resonant of the messy empirical structures that adorn large-scale atlases and maps, yet this is not to say that they are devoid of structure. The type of analysis presented in this book by Longley and Mesev (Chapter 9) emphasizes the scaling relations that characterize the structure and form of settlements, and suggests how fractal geometry may be used to detect systematic structured irregularity. Having re-established a basis for linking form to function, it should become possible to appraise the extent to which different urban forms are 'sustainable', according to a wide range of criteria. Urban remote sensing also offers the prospect of extending two-dimensional analysis of form into three dimensions, for example through the use of LiDAR and interferometric SAR (Synthetic Aperture Radar) (Grey and Luckman 1999). It also offers the prospect of developing much better models of the dynamics of urban growth and change, using approaches like those developed by Batty and Howes (Chapter 10, this volume).

13.6 Conclusion

Pesaresi and Bianchin (Chapter 4, this volume) make the point that the Italian 'urbanist' (planner-architect) tradition carries forward a rich understanding of urban structure and of the relations between form and function, but that this has yet to be related to the fast-developing field of urban remote sensing. The latter already offers reasonably detailed urban classifications, yet is blind to pattern, doggedly reductionist in its operation and largely oblivious to urban function. It remains to be seen whether the metaphors that 'urbanists', geographers and planners use to structure our understanding of urban form and function (Johnston 1999) can be used to inform new methods of incorporating spatial syntax and ancillary data sources.

The outcome of research will determine whether urban remote sensing is to fulfill only a technical role in corroborating and updating other data sources, or whether it might fulfill a more central role in terms of quality of life studies, and data-rich modelling of form and function. Some of the 'softer' issues of urban classification discussed here also have hard-edged resource implications (Weber, Chapter 8 this volume). Urban remote sensing opens up enticing new prospects for comparing cities in terms of functional interrelationships and indicators of well-being, and the integration of the study of spatial forms with an understanding of the social, economic, cultural and political dimensions that led to their creation. Ironically, as the quantity and quality of remotely-sensed data improves, so our sense of certainty and confidence in our urban classifications appears to diminish. A healthy dose of scepticism may also make an important contribution as we move from trying to understand data to understanding the urban systems to which they pertain.

13.7 References

Atkinson, P., and Martin, D. (editors), 2000, *GIS and Geocomputation* (London: Taylor and Francis).

Atkinson, P., and Tate, N. (editors), 1999, *Advances in Remote Sensing and GIS Analysis* (Chichester: John Wiley and Sons).

Barnsley, M., 1999, Digital remotely-sensed data and their characteristics, In *Geographical Information Systems: Principles, Techniques, Management and Applications*, edited by P. A. Longley, M. F. Goodchild, D. J. Maguire, and D. W. Rhind (New York: John Wiley), pp. 451–466.

Barr, S. L., and Barnsley, M. J., 1999, A syntactic pattern recognition paradigm for the derivation of second-order thematic information from remotely-sensed images, In *Advances in Remote Sensing and GIS Analysis*, edited by P. Atkinson, and N. Tate (Chichester: John Wiley and Sons), pp. 167–184.

Batty, M., and Longley, P. A., 1994, *Fractal cities: A geometry of form and function* (London: Academic Press).

Batty, M., and Longley, P. A., 1996, Analysis, modelling, forecasting and GIS technology, In *Spatial Analysis: Modelling in a GIS Environment*, edited by P. A. Longley, and M. Batty (New York: John Wiley), pp. 1–15.

Berry, B. J. L., 1973, *The Human Consequences of Urbanisation: Divergent Paths in the Urban Experience of the Twentieth Century* (London: Macmillan).

Binaghi, E., Brivio, P. A., Ghezzi, P., Rampini, A., and Zilioli, E., 1996, A hybrid approach to fuzzy land cover mapping. *Pattern Recognition Letters*, **17**, 1399–1410.

Blamire, P., 1996, The influence of relative sample size in training artificial neural networks. *International Journal of Remote Sensing*, **17**, 223–230.

Cho, S.-B., and Kim, J. H., 1995, Combining multiple neural networks by fuzzy integral for robust classification. *IEEE Transactions on Systems, Man and Cybernetics*, **25**, 380–384.

Curran, P. J., Milton, E. J., Atkinson, P. M., and Foody, G. M., 1998, Remote sensing: from data to understanding, In *Geocomputation: a Primer*, edited by P. A. Longley, S. M. Brooks, W. D. Macmillan, and R. McDonnell (Chichester: John Wiley), pp. 33–59.

Fisher, P., 1999, Models of uncertainty in spatial data, In *Geographical Information Systems: Principles, Techniques, Management and Applications*, edited by P. A. Longley, M. F. Goodchild, D. J. Maguire, and D. W. Rhind (New York: John Wiley), pp. 191–205.

Fisher, P. F., and Pathirana, C., 1990, The evaluation of fuzzy membership of land cover classes in the suburban zone. *Remote Sensing of Environment*, **34**, 121–132.

Foody, G. M., 1996, Approaches for the production and evaluation of fuzzy land cover classifications from remotely-sensed data. *International Journal of Remote Sensing*, **17**, 1317–1340.

Foody, G. M., 1997, Fully fuzzy supervised classification of land cover from remotely sensed imagery with an artificial neural network. *Neural Computing and Applications*, **5**, 238–247.

Foody, G. M., and Arora, M. K., 1997, An evaluation of some factors affecting the accuracy of classification by an artificial neural network. *International Journal of Remote Sensing*, **18**, 799–810.

Goodchild, M. F., and Longley, P. A., 1999, The future of GIS and spatial analysis, In *Geographical Information Systems: Principles, Techniques, Management and Applications*, edited by P. A. Longley, M. F. Goodchild, D. J. Mcguire, and D. W. Rhind (New York: John Wiley), pp. 567–580.

Grey, W., and Luckman, A., 1999, Using SAR interferometric phase coherence to monitor urban change, In *RSS'99: Earth Observation: From Data to Information* (Nottingham: Remote Sensing Society), pp. 453–462.

Guptill, S., 1999, Metadata and data catalogues, In *Geographical Information Systems: Principles, Techniques, Management and Applications*, edited by P. A. Longley, M. F. Goodchild, D. J. Maguire, and D. W. Rhind (New York: John Wiley), pp. 677–692.

Haggett, P., Cliff, A. D., and Frey, A., 1977, *Locational Analysis in Human Geography* (London: Edward Arnold).

He, H., Collet, C., and Spicher, M., 1999, Evaluation and comparison of two fuzzy classifiers for multi-spectral imagery analysis, In *IGARSS99* (Hamburg: IEEE).

Johnston, R., 1999, Geography and GIS, In *Geographical Information Systems: Principles, Techniques, Management and Applications*, edited by P. A. Longley, M. F. Goodchild, D. J. Maguire, and D. W. Rhind (New York: John Wiley), pp. 39–47.

Lee, H.-M., and Lee-Kwang, H., 1994, Information aggregating networks based on extended Sugeno's fuzzy integral, In *IEEE/Nagoya University World Wisepersons Workshop*, IEEE (Springer-Verlag), pp. 56–66.

Longley, P., 1998, GIS and the development of digital urban infrastructure. *Environment and Planning B*, **25**, 53–56.

Longley, P., Batty, M., Shepherd, J., and Sadler, G., 1992, Do green belts change the shape of urban areas? A preliminary analysis of the settlement geography of South East England. *Regional Studies*, **26**, 437–452.

Longley, P., Brooks, S., Macmillan, W., and McDonnell, R. (editors), 1998, *Geocomputation: A Primer* (Chichester: John Wiley).

Longley, P., and Mesev, T., 1999, On the measurement and generalisation of urban form. *Environment and Planning A*, **32**, 473–488.

Longley, P. A., and Harris, R. J., 1999, Towards a new digital data infrastructure for urban analysis and modelling. *Environment and Planning B*, **26**, 855–878.

Mather, P., 1999, *Computer Processing of Remotely-Sensed Images*, second edition (Chichester: John Wiley and Sons).

OPCS, 1984, *Key Statistics for Urban Areas*, Office of Population Census and Surveys (London: HMSO).

Openshaw, S., 1984, *The Modifiable Areal Unit Problem*, volume 38 of *Concepts and Techniques in Modern Geography* (Norwich: GeoBooks).

Paola, J., and Schowengerdt, R., 1997, The effect of neural network structure on multispectral land-use/land-cover classification. *Photogrammetric Engineering and Remote Sensing*, 63, 535–544.

Rhind, D., 1997, *Framework for the world* (New York: John Wiley).

Timms, D., 1971, *The Urban Mosaic: Towards a Theory of Residential Differentiation* (Cambridge: Cambridge University Press).

Wu, F., 2000, A parameterised urban cellular model combining spontaneous and self-organising growth, In *GIS and Geocomputation*, edited by P. Atkinson, and D. Martin (London: Taylor and Francis).

Zang, J., and Foody, G. M., 1998, A fuzzy classification of sub-urban land cover from remotely sensed imagery. *International Journal of Remote Sensing*, 19, 2721–2238.

Index

Printed and bound by CPI Group (UK) Ltd, Croydon, CR0 4YY

23/10/2024

01778232-0007